Archaeology and Clays

Edited by

Isabelle C. Druc

BAR International Series 942
2001

Published in 2019 by
BAR Publishing, Oxford

BAR International Series 942

Archaeology and Clays

ISBN 9781841711751 paperback
ISBN 9781407352909 e-book

DOI https://doi.org/10.30861/9781841711751

A catalogue record for this book is available from the British Library

This book is available at www.barpublishing.com

BAR Publishing is the trading name of British Archaeological Reports (Oxford) Ltd. British Archaeological Reports was first incorporated in 1974 to publish the BAR Series, International and British. In 1992 Hadrian Books Ltd became part of the BAR group. This volume was originally published by John and Erica Hedges in conjunction with British Archaeological Reports (Oxford) Ltd / Hadrian Books Ltd, the Series principal publisher, in 2001. This present volume is published by BAR Publishing, 2019.

BAR
PUBLISHING

BAR titles are available from:

BAR Publishing
122 Banbury Rd, Oxford, OX2 7BP, UK
EMAIL info@barpublishing.com
PHONE +44 (0)1865 310431
FAX +44 (0)1865 316916
www.barpublishing.com

CONTENTS

INTRODUCTION
ARCHAEOLOGY AND CLAYS

Isabelle C. Druc
Department of Anthropology, University of Wisconsin-Madison

One of the main problems facing archaeologists interested in ceramic provenance and technology is to determine the type and origin of the materials and resource areas used by ancient potters. However, information on regional clay variability and local potential for ceramic production is usually sparse. Geologists and archaeologists have different objectives and ways of looking at a landscape, rock or soil. Furthermore, geological maps rarely mention non-industrial clay deposits or materials of interest to potters and archaeologists. The symposium from which this volume is inspired was meant to bring geologists, clay mineralogists and archaeologists together, to discuss these issues.

The symposium was held during the 37th annual Clay Minerals Society meeting, at the Water Tower Campus, Loyola University in Chicago, in June 2000. The idea was to get as wide a range of analyses and interpretations as possible concerning clays and ceramics. Having the symposium held during the Clay Minerals Society meeting allowed for the clay mineralogists to get interested in the problems particular to archaeological materials and archaeo-ceramicists to explore the clay side of their material. Mineral identification and characterization, material sourcing, regional survey of available resources for potters, and estimation of firing temperatures were among the main topics brought up. Contrary to archaeometry symposiums where the emphasis is on the characterization of ceramic materials using physico-chemical techniques and bulk analyses (i.e. INAA and EDXRF, see the abbreviation list at the end of the chapter), the present symposium focused on the material in its mineralogical aspect, both at the plastic (clay) and non plastic (temper) phases to resolve archaeological problems.

The techniques used in the studies presented in this volume range from optical to chemical, with emphasis on petrography, spectroscopy and x-ray diffraction. These techniques allow one to characterize the different components or phases in a ceramic, to study their mutual relationships or influence upon the properties of the paste, their genetic origin, or estimation of firing temperatures. In general, a variety of complementary techniques are applied to correlate the results and strengthen data interpretation. Such analyses lead to the determination of provenance, understanding ceramic technology and technological changes, or distinguishing ware types. Several chapters in this volume are concerned with the question of provenance directly or indirectly (Shriner and Murray, Bossiere and Frère, Druc, Miksa). Characterizing the different local paste types from a site may point to the existence of different production centers, a shift in resource locations, or a change in the technology (Shriner and Murray). Technology and organization of production are also investigated (Friedman, Eygun, Velde and Bouchain, Druc). Mineral and chemical characterization can link ceramic typologies based on style, form or surface appearance to clays and pastes, eventually leading to identification of social groups and populations (Cecil). The analysis of the clays and non plastic inclusions also allows estimation of firing temperatures (Bruni et al.), as melting or new minerals form when certain temperatures are reached. In addition, laboratory experiments, analysis of comparative material and geological survey are usually conducted to ground the interpretation of ceramic analysis, as exemplified in the chapters of this volume.

One aspect which is necessary to stress is the complex character and variability of the material archaeologists are dealing with. Unlike industrial ceramics, archaeological ceramics are usually composed of clays that contain variable amounts of inclusions, natural temper grains, organic material, and other 'impurities' (Velde and Druc 1999). These clays, fine or coarse, are used alone or mixed with other clays, soils, sand, crushed rock, bone, shell, or other materials. Any intentional addition to the clay base is called temper in the archaeological literature, and one important objective in ceramic studies is to determine the different components used by the potter. Ethnographic studies and paste analysis offer an idea of the wide range of materials that are used in traditional production. When mixing soils, grinding source materials, or withdrawing the coarser fraction of a sediment by levigation or other methods, the potter's objective is to produce a clay body or paste that, when water is added, will have the correct working plasticity to work with and render the type of paste suitable for the ware he or she wishes to produce. The potting technique, surface treatment and firing process also influence the properties of the finished product, such as its ability to withstand shocks and temperature changes, or its porosity. These properties have functional and economic consequences, while the quality or type of the wares may determine their use in different social networks and occasions. Environmental constraint is important, but the human factor in the variability observed in ceramics is tremendous. This is an aspect geologists do not have to contend with when analyzing sediments and clays. Hence the importance of an interdisciplinary dialogue.

Because of the background of many contributors and of the orientation of their research and methodological beliefs, the studies in this volume emphasize the identification of the different constituents in the ceramic paste. These constituents are the clay particles and non plastic inclusions, natural or present as part of a second or third ingredient added by the potter to the clay. By identifying the paste constituents and characterizing their mineral and chemical compositions, much can be learned about the provenance of the materials and the wares, and information can be gathered on ancient ceramic technology. The identification of ware provenance is based on the premise that potters will preferably use the materials available locally or in their region. It is assumed that the majority of the ceramics found on one archaeological site would be local. However, in many cases, trade wares constitute a certain percentage *per se* or as containers for non local goods that were brought to the site. Identifying the provenance of the non local ceramics yields clues about politics, economy and cultural interactions between sites or regions. It is a way to expand our knowledge about ancient cultures and civilizations. This requires one not only to study the materials but also to have a good knowledge of the local geology, of the clays and other resources available, and of the archaeological and cultural contexts of the samples analyzed.

Because of the inherent complexity of the material, of its often multiple origins (if more than one material is combined), of the transformations introduced by the potter when preparing the paste or by firing the wares, it is helpful to study the clays and non plastics before engaging in bulk analysis. The latter refers to chemical analysis of the whole paste (matrix and inclusions) crushed into powder samples. Although important information on chemical composition is acquired, precious information on ware technology is lost by the preparation process of the samples. The interpretation of the chemical data is also significantly eased when mineralogical information is present. Not one technique, however, will answer all the questions archaeologists are concerned with. Accordingly, analysis techniques and disciplines must combine to yield a comprehensive picture of ware production and provenance, technology and human behavior. To reach this goal, analysts, mineralogists, and archaeologists must work together.

The interdisciplinary scope of the studies presented here is enhanced by the dual background of many contributors (with training in anthropology and different aspects of ceramic analysis), or as the result of team efforts, working in close partnership and involving scholars specialized in material sciences, chemistry and geology, and archaeologists, art historians and classical academics. This insures a better understanding of the problem to be resolved. This dual approach and flexible vision is reflected in the chapters presented in this volume.

This interdisciplinary trend has been advocated since the mid eighties, in conjunction with an urge to return to a more global view of production and an understanding of man and society (Matson 1984; Kolb 1988). Although the anthropological and social perspective is not always explicitly expressed here, the level of data interpretation goes beyond the analysis exercise to answer questions raised by archaeologists and better understand the potter and his work. The international panel of contributors in this volume also offers a broad perspective of clay and ceramic analysis from different geographical areas, in both the Old and New Worlds.

Christine M. Shriner and Haydn H. Murray work toward understanding the ceramic technology at the Greek site of Lerna during the third millennium BC (Chapter 1). Shriner is a classical archaeologist who has training in geology. Murray, a clay mineralogist, is professor emeritus of geology at Indiana University. In addition, he is head of the Applied Clay Mineralogy Laboratory in the Department of Geological Sciences at Indiana University. Stylistic analysis and compositional data led Shriner to propose a radically new understanding of ceramic production at Lerna, at a time when new ceramic shapes and decoration occur. Electron microprobe and XRD analyses of local clay sources and Lernean ceramics allowed Shriner and Murray to ascertain the local character of the production for the wares with metamorphic paste, thought once to be imports. Furthermore, they identified a shift in clay sources through time and a change of technology to adapt to the new clay body.

Elizabeth Friedman's study of Anatolian metallic ware (Chapter 2) investigates the material characteristics of these hard and thin ceramics, formerly believed to be fired at high temperature due to the metallic ring they make when struck. In order to assess the composition and firing temperature of these wares, Friedman, an archaeologist actively engaged in ceramic analysis, conducted SEM/EDS studies, completed with optical microscopy, XRD, and INAA studies. The result of this combined analytical approach shows that the metallic aspect of the ware is not due to vitrification but to particular chemical and mineralogical attributes.

Chapters 3 and 4 deal with the analysis of Etruscan ceramics in Italy. The Italian team of Silvia Bruni and colleagues, composed of members from departments of Inorganic Chemistry, Sciences of the Antiquity, and Earth Sciences from Milan and Padua, exploit the potential of Fourier-Transform infrared spectroscopy and near-infrared diffuse reflectance spectroscopy, rarely used to characterized archaeological ceramics. The former method (FTIR) specifies the mineralogical phases in the paste, to later detail them by X-ray powder diffraction. The latter method (NIR) allows approximation of the firing temperature. In addition, diffuse reflectance in the visible range and Mössbauer spectroscopy yield comparative assessments.

The second study of Etruscan ceramics (Chapter 4) by Gérard Bossière and Dominique Frère is focused on the use of elemental x-ray spectroscopy by scanning electron microscopy (EDX-SEM) to determine the signature expected in wares from different regions. This technique allows characterization of the composition of specific grains or areas of the paste by aiming the electron beam at them, yielding a 'spot analysis,' in contrast to a bulk analysis. Bossière, a French geologist, came into contact with archaeology when a student of Art History asked him to identify minerals in Etruscan ceramics. As a specialist in metamorphic rocks, Bossière approaches ceramics as the result of contact metamorphism.

One last study focusing on the Italian Peninsula is presented in Chapter 5 by Guilmine Eygun, an archaeologist with training in ceramic analysis. Eygun presents a technological study of Neolithic pottery of the Tavoliere Plain in southeast Italy. The characterization of the different ceramic pastes is based on petrography and X-ray diffraction, with comparative studies of local clay deposits. The results are interpreted in light of the regional geology, and the author goes beyond the analysis to relate the results to the potters' behavior and functional strategies.

Archaeologist Isabelle Bouchain and clay mineralogist Bruce Velde offer an image analysis view of Gaulo-Roman ceramics from central France (Chapter 6). The idea is to isolate granulometric patterns in thin sections using an imaging system and looking at grain distribution patterns. The authors then relate these patterns to material acquisition behaviors, determining the use of simple or multiple sediments, and to methods of manufacture and ceramic production.

Beth Miksa's strong background in ceramic petrography and southwest archaeology offers a very systematic, mineralogical approach to the study of ceramic provenance (Chapter 7). She applies granulometric, textural and compositional criteria to distinguish the different plastic and aplastic components of a ceramic paste and to determine their genetic origin. This prior knowledge is essential to sample correctly the ceramics for bulk chemical analysis and to interpret the chemical data. Furthermore, her study of Hohokam buff ware lead her to conduct experimental studies with crushed schist to distinguish between naturally occurring schist within the clay and added material. She advocates the study of mineral and clay resources available to the potters by building compositional reference charts based on point counting analysis of sand and silt samples to portray the variability of the sources. The study of the silt and sand components of the clay deposits allows her to distinguish natural versus added temper in the ceramics, an important topic in ceramic analysis.

Isabelle Druc approaches the question of paste characterization and ware provenance based on ethnoarchaeological data. As an archaeologist trained in materials sciences, she seeks to understand past productions and potters' behavior in an interdisciplinary way, combining the analysis of raw materials and resources to the study of acquisition strategies and preparation procedures. Prior studies allow her to present a diachronic perspective of the materials used for ceramic production in the north central Andes. The use of crushed slate or shale as temper by a number of neighboring communities and the presence of similar material in earlier ware prompted the characterization of this tempering material on a regional scale. The objective was to identify compositional differences to allow for a correct identification of provenance. However, acquisition strategies implying the use of the same resources by different potting villages blur the picture at the local level and suggest the use of complex explanatory models for ware production in the past.

Finally, Leslie Cecil triangulates paste composition with style and technological characteristics to highlight social identities among Péten Postclassic Maya potters (Chapter 9). She draws on petrography, XRD and strong acid-extraction ICPS to characterize different technological styles of Petén Postclassic slipped ware. Cecil is an anthropologically trained archaeologist who specializes in archaeological chemistry. The anthropological background of the author lends to an extended interpretation of the analysis results, tackling identity issues, ideology and social contexts.

As can be seen, this volume emphasizes a diversified approach to the analysis of clay artifacts, combining mineralogy, archaeology and chemistry in a true interdisciplinary way. The importance of understanding the local geology and mineral resources is also stressed. It is not sufficient to have geologists identifying lists of minerals and archaeologists trying to interpret the results. To engage in constructive discussion and sound explanation of ancient technologies and societies, scholars should know enough of each other's disciplines to understand the analyses done and the associated research objectives.

Abbreviations and List of Analysis Techniques Mentioned in this Volume

Below is a list of abbreviations and analysis techniques mentioned in this volume, followed by a short descriptive. A full description of these techniques can be found in numerous publications on ceramic analysis (e.g. Bishop et al. 1982; Goffer 1980; Rice 1987; Velde and Druc 1999).

Optical techniques

Petrography. The mineral analysis is performed with a petrographic microscope using 30 micron thick thin sections. The optical characteristics seen by transmitted light help identify the type, size, angularity, and aspect of minerals and grains in the paste. The resolution of this microscope is however not powerful enough to study clay-size minerals. These can be studied with a scanning electron microscope (SEM).

Spectroscopic techniques

EDS/SEM This technique combines energy dispersive spectroscopy (EDS) and optical microscopy. The EDS device is coupled to a scanning electron microscope (SEM). This allows selecting the area or grain for chemical analysis. The technique measures the secondary x-rays emitted upon the interaction with the electrons in a way similar to XRF analysis using energy dispersive spectroscopy. It allows for the identification of major and minor elements on a semi-quantitative basis (with internal standards).

EMPA Electron Microprobe Analysis. As for EDS/SEM, the microprobe is coupled to an electron microscope. However, the technique uses wavelength dispersive spectroscopy to give composition by oxides of individual minerals on a quantitative basis.

FTIR Fourier-Transform infrared spectroscopy. The method helps determine mineralogical phases. It measures the absorption of radiation in the infrared range at the molecular level.

ICPS Inductively coupled plasma emission spectroscopy. The technique allows to determine the chemical composition of a sample, using an atomic-emission or atomic-absorption spectrometer for the analysis. The preparation method requires the dissolution of the sample in a dilution. The strong acid-extraction preparation method is the method employed in Chapter 9.

Mössbauer spectroscopy. The technique helps estimate firing temperature based on the determination of the oxidation state of iron.

NIR Near-infrared diffuse reflectance spectroscopy. The technique helps estimate the firing temperature by measuring the amount of absorbed water in a sample.

XRD X-ray diffraction. A spectroscopic technique to determine the identity of crystalline materials, often used for the identification of raw clays. The firing of the ceramic may destroy the crystal lattice if temperatures are high enough, prohibiting the use of this technique for melted bodies. However the method is very useful to identify newly formed, micro-crystalline minerals produced during firing.

Bulk physico-chemical techniques

INAA Instrumental neutron activation analysis. A bulk chemical analysis to study the elemental composition of a paste, mainly its rare, trace and minor elements. The technique induces the excitation of the atomic nucleus, producing radioisotopes, whose decay time and radiation are characteristic of the elements present in the sample.

XRF X-ray fluorescence. A bulk chemical analysis to study the elemental composition of a paste, mainly its minor and major elements. XRF yields a quantitative analysis by emission spectroscopy of secondary x-rays. A beam of x-rays hits the electrons of the inner shell, which induces their excitation and ejection from their original orbit. An electron from the upper orbits fills the empty space. The change of energy level induces the release of secondary rays or fluorescence, characteristic of each atom. The technique measures the energy or wavelength of the x-rays. EDXRF is energy dispersive x-ray fluorescence.

Acknowledgments

Bruce Velde has been the initial instigator of the symposium on Archaeology and Clays presented at the Clay Minerals Society meeting, which I co-organized with him. For this and his constant support, may he be heartfully thanked. The papers in this volume have been submitted to two peer reviews and benefited from comments subsequent to the symposium. I want to thank the persons who helped me in this task, in particular Bruce Velde, Haydn H. Murray, Christine M. Shriner, Silvia Bruni and Guilmine Eygun.

References

Bishop, R. L., Rands, R.L., and Holley, G.R. 1982. Ceramic compositional analysis in archaeological perspective. *Advances in Archaeological Method and Theory*, 5: 275-330.

Goffer, Z. 1980. *Archaeological chemistry. A source book on the applications of chemistry to archaeology*. New York: John Wiley and Sons.

Kolb, C. 1988. Ceramic ecology in retrospect: A critical review of methodology and results. In *Ceramic Ecoloy Revisited 1987*. ed. C.C. Kolb. Oxford: British Archaeological Reports International Series 436(i), pp. 261-375.

Matson, F. 1984. Ceramics and man reconsidered with some thoughts for the future. In *The many dimensions of pottery: Ceramics in archaeology and anthropology*, eds. S.E. van der Leeuw and A.C. Pritchard. Amsterdam: Universiteit van Amsterdam, pp. 25-50.

Rice, P. 1987. *Ceramic analysis. A source book*. Chicago: University of Chicago Press.

Velde, B., and Druc, I. C. 1999. *Archaeological ceramic materials. Origin and utilization*. Berlin Heidelberg: Springer Verlag.

List of Contributors

Artioli G.	Dipartimento di Scienze della Terra, Università degli Studi di Milano, Via Mangiagalli 34, 20133 Milano, Italy
Bagnasco Gianni G.	Dipartimento di Scienze dell'Antichità, Università degli Studi di Milano, Via Festa del Perdono 7, 20122 Milano, Italy
Bonghi Jovino M.	Dipartimento di Scienze dell'Antichità, Università degli Studi di Milano, Via Festa del Perdono 7, 20122 Milano Italy
Bossière G.	Laboratoire de Planètologie et Géodynamique (FRE CNRS 2129) Faculté des Sciences et des Techniques de Nantes, 2 rue de la Houssinière, BP 92208, 44322 Nantes Cedex 3, France (UMR CNRS 6112) Gerard.Bossiere@chimie.univ-nantes.fr WEBsite :http://www.sciences.univ-nantes.fr/scnat/geologie/ea2156te
Bouchain I.	UMR Archéologie et Sciences de l'Antiquité, EP 1630, Université de Nanterre, 92023 Paris, France
Bruni S.	Dipartimento di Chimica Inorganica, Metallorganica e Analitica, Universita degli Studi di Milano, Via Venezian 21, 20133 Milano, Italy sbruni@csmtbo.mi.cnr.it
Cariati F.	Dipartimento di Chimica Inorganica, Metallorganica e Analitica, Universita degli Studi di Milano, Via Venezian 21, 20133 Milano, Italy
Cecil L.	Department of Anthropology, Mailcode 4502, Southern Illinois University at Carbondale, Carbondale, IL 62901 hunapu@siu.edu
Druc I. C.	Department of Anthropology, 5240 Social Science Building, 1180 Observatory Drive, Madison WI 53706-1393 USA icdruc@facstaff.wisc.edu
Eygun G.	Département d'anthropologie, Université de Montréal, C.P. 6128 Succ. Centre Ville, Montréal, Québec H3C 3J7, Canada g_eygun@hotmail.com
Frère D.	Université de Bretagne Sud, UFR d'Histoire, 4 rue Jean Zay, 56325 Lorient Cedex, France (ETUMR CNRS 126-4) dom.frere@infonie.fr
Friedman, E. S.	Department of Near Eastern Languages and Civilizations, The University of Chicago, 1155 E. 58th St. Chicago, IL 60637 esf1@midway.uchicago.edu
Miksa E. J.	Desert Archaeology, Inc., 3975 N. Tucson Blvd., Tucson AZ 85716 USA emiksa@desert.com
Murray H. H.	Department of Geological Sciences, Indiana University, 1005 E. 10th Street, Bloomington, IN 47405 Murray@Indiana.edu
Russo U.	Dipartimento di Chimica Inorganica, Metallorganica e Analitica, Università degli Studi di Padova, Via Loredan 4, 35100 Padova, Italy
Shriner C. M.	Department of Geological Sciences, Indiana University, 1005 E. 10th Street, Bloomington, IN 47405 cShriner@Indiana.edu
Velde B.	Laboratoire de Géologie, École Normale Supérieure, 24 rue Lhomond, 75231 Paris Cedex 05 papa@geologie.ens.fr

CHAPTER 1

EXPLAINING SUDDEN CERAMIC CHANGE AT EARLY HELLADIC LERNA: A TECHNOLOGICAL PARADIGM

Christine M. Shriner
Department of Geological Sciences and Department of Classical Studies, Indiana University

Haydn H. Murray
Department of Geological Sciences, Indiana University

Dedicated to the memory of Richard E. Peeler, Master Potter

This research examines assumptions about ceramic change at Early Helladic Lerna in light of integrated analytical studies on both ceramic artifacts and potential local clay-rich sediments. Petrographic (PE), electron microprobe (EMPA) and X-ray diffraction (XRD) projects were combined to evaluate the extent of local ceramic raw material availability and the technological capabilities for ceramic production. Whereas petrography and electron microprobe are used in this methodology to predict and quantitatively confirm local versus non-local ceramic provenance, X-ray diffraction is used to reconstruct aspects of the ceramic technology employed at Lerna during the cultural periods III (ca. 2650-2150 BC) and IV (ca. 2150-2000 BC). Changes in the ceramic technology are used as evidence for better understanding the profound cultural changes at the Lerna III/IV boundary.

INTRODUCTION

Nature of the Problem

Lerna, a prehistoric site in the Argolid, Greece, displays a well-stratified sequence from the Final Neolithic (ca. 4000-3000 BC) through the Early Helladic III (2150-2000 BC) period (Figure 1). Significant ceramic technological changes occurred at Lerna over the course of the Third Millennium; however, previous ceramic studies at Lerna for this time period have focused primarily on changes in stylistic features such as shape and surface decoration as indicators of social and economic change (Rutter 1995; Wiencke 2000).

Despite the fact that technological change is mentioned in these studies the technological ceramic evidence has not been interpreted culturally by the same standards as the stylistic evidence and has more often been relegated to reports isolated in the appendices of publications. Following the lead of Matson (1965), Jones (1986, 1993) and Whitbread (1995), significant ceramic technological changes at Lerna over the course of the Third Millennium cannot be excluded from discussion as one of many "events" involved in a major shift in human behavior.

Results of Ceramic Provenancing (PE and EMPA) Projects

The results of the original petrographic analyses for Lerna III and IV (Betancourt and Myer 2000; Betancourt et al. 1995; Vaughan et al. 1995) documented three major fabric groups in the sherd material: (1) fine calcareous; (2) biogenic limestone; and (3) disparate metamorphic. The mineral constituents and proportions that characterize the PE fabric groups are a reflection of the natural sediments of the local Xerias River system, i.e. a raw material continuum (Vaughan et al. 1995: 693-710).

Using local clay-rich sediments as a standard of reference, EMPA compositional data were compared with the Lernean sherd material (Shriner and Dorais 1999). Muscovite chemical compositions in reference and sherd materials overlapped. In addition two minerals, chloritoid and tourmaline, which are present in the reference samples but had previously been unidentified in the sherds, were analyzed in the sherd thin-sections. Identical chloritoid, tourmaline and muscovite compositions in both source materials and sherds presented a quantitative geomarker or fingerprint for local ceramic provenancing. Comparative microprobe analyses substantiated a local source for virtually all of the ceramic classes spanning cultural periods III and IV. This research and the sample to date suggested that 95% of the Lerna III and IV ceramics were local, with only a few exotics appearing in the sample.

The verification of local clay compositions used at Lerna and the variability within those clay sources are crucial evidence in the reconstruction of the ceramic technology employed at Lerna during Lerna III and IV. Once local provenancing was established, other technological parameters could be identified through clay mineralogical analysis, including XRD and physical property testing, of ceramic artifacts and local clay-rich sediments. The present study deals specifically with the results of the Lerna XRD project and discusses alternative technological explanations for some of the profound stylistic changes in the Lerna III and IV assemblages.

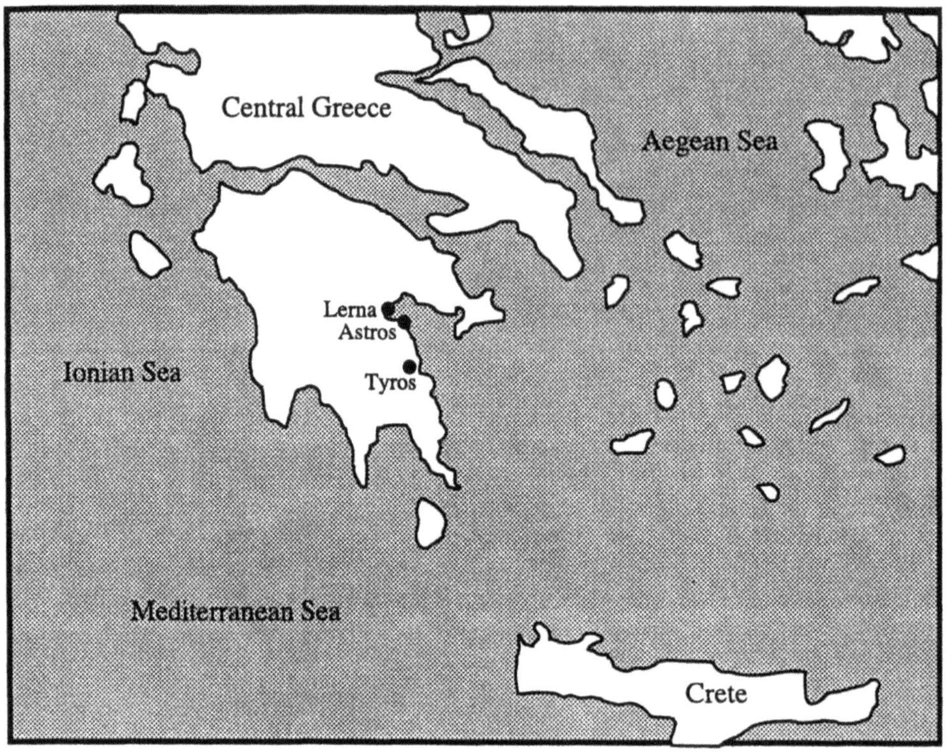

Figure 1. Generalized map of Greece and the surrounding area showing the location of Lerna.

Technological Premise

Clay matrix compositional data for Lerna III and Lerna IV sherds, which represent the local petrographic fabrics (Disparate Metamorphic, Biogenic Limestone, and Fine Calcareous) for each period, were plotted using a diagram adopted from Philpots (1994). Philpots suggested that a hypothetical clay composition for archaeological ceramic material could be estimated using EMPA analysis on clay matrix. Philpots used elemental compositions for illite and smectite derived from Weaver (1989) as parameters and plotted octahedral (Fe, Mg) cations against tetrahedral (Al) and interlayer cations (K, Na, Ca). The plots fall between Philpot's suggested parameters for muscovite/illite and smectite. Two discrete groupings are recognized in the diagram (Figure 2). The Lerna III sherds have a more muscovite/illite-rich clay composition. The Lerna IV samples display a composition consistent with randomly-layered illite-smectite clays (Deer et al. 1992).

As a point of departure for this study the present authors proposed that the Lernean potters made a conscious decision to change and/or utilize a different clay source within their local metamorphic clays during the Lerna III and IV periods. Representative ceramic assemblages for Lerna III/Phase Late A and Lerna III/Phase Early C were chosen from a new 200 count technological sample along with a Lerna IV/Phase 1 assemblage from the original stylistic sample for comparative XRD analysis with the local clay-rich sediments.

ARCHAEOLOGICAL PROBLEM

Lerna III/IV Boundary

Caskey's excavations at Lerna revealed that the site was occupied in the Early, Middle and Late Neolithic (ca. 6000-3000 BC) periods (Lerna levels I and II), after which there appeared to have been a hiatus or at least a much reduced occupation in the Early Helladic I period. In the Early Helladic II (ca. 2650-2150 BC) period (Lerna level III) there arose at Lerna a walled settlement dominated by large rectangular structures such as Building BG and the "House of Tiles". Signs of developing social complexity, e.g. caches of seal impressions, were found in the "House of Tiles".

Pottery from EH II deposits consisted of an unburnished fine ware with a black to brown to red slip/paint (EH Urfirnis) and a fine burnished ware with a light slip/paint (Yellow Mottled). Shapes in these wares included: saucers, sauceboats, basins, bowls, askoi and beaked jugs. A coarse, dark-surfaced and unburnished ware was used for pithoi, hearths, large cooking bowls, stands and baking pans (1960: 289-293).

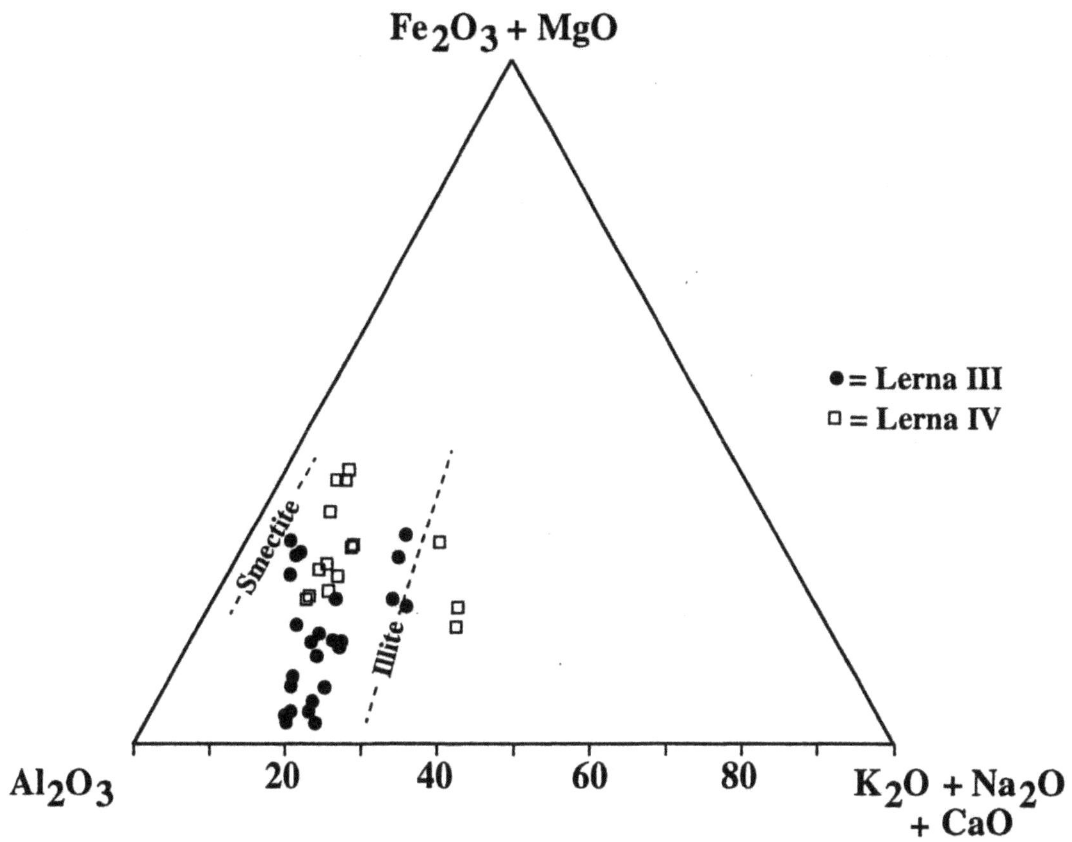

Figure 2. Fe_2O_3 + MgO – Al_2O_3 – K_2O + Na_2O + CaO compositions of the matrix clays of the Lerna sherds (after Philpots 1994). Lerna III samples are richer in illite than Lerna IV sherds, which appear to have been made from a randomly-layered illite-smectite clay.

Caskey observed that shortly after the erection of the EH II "House of Tiles" structure, the site was burnt and the subsequent Early Helladic III (ca. 2150-2000 BC) community (Lerna level IV) was unwalled and consisted of smaller structures, apsidal and oblong in plan. Certain ceramic shapes such as the sauceboat were no longer present. New shapes such as tankards, one-handled cups, two-handled bowls and ouzo cups were introduced. New pottery classes included Dark-on-Light Patterned, Dark Burnished and some early Gray Minyan (1960: 293-297).

The cultural break or "event", observed by earlier researchers as a destruction level at various other sites, occurs at Lerna between EH II (Lerna level III) and EH III (Lerna level IV), not between EH III and the Middle Helladic (MH). As there was no adequate temporal correlation at this point, Caskey made Lerna the standard. In his 1960 analysis he revised the date for the destruction levels observed by others at various sites in the Argolido-Corinthia study area by identifying the dates of those destructions in terms of their place within the Lerna sequence. He proposed that Tiryns, Asine, Zygouries, Ayios Kosmas and perhaps Corinth were destroyed at the same time as Lerna by an invading group. Korakou and Eutresis, he suggested, survived the destruction at the end of the EH II and were destroyed at the end of the EH III period (1960: 301-303).

To strengthen his hypothesis Caskey marshaled together not only stratigraphical evidence - e.g. destruction layers and periods of abandonment - but evidence of abrupt changes in the material assemblages (i.e. architecture, ceramics, lithics and burial customs) at Lerna, and by comparison, at other sites in the region (1960: 289-297). Caskey isolated three presumably "non-local" and innovative classes in the Lerna IV pottery assemblage: (1) Slipped and Burnished; (2) Moderately Fine Burnished; and (3) Early Gray Minyan (1960: 295-297). The appearance of these "non-local" ceramic classes at Lerna played a prominent role as supporting evidence in Caskey's invasion theory. Unfortunately, Caskey's ceramic analysis for the Lerna III/IV boundary was preliminary and not completed by him before his death.

Recently, Rutter has developed a classification system for the Lerna IV ceramic corpus and offered a modified explanatory model for the abrupt ceramic changes between Lerna III and IV (1995). The results of Rutter's stylistic analysis for Lerna IV indicate that the earliest EH III (Lerna IV) pottery from Lerna was a mixture of indigenous elements (i.e. continuities from Lerna III) with a series of intrusive stylistic elements, derived from western Anatolian and Cycladic models which coalesced in central Greece before moving south into the Peloponnese (1995:477, 649).

Caskey's original "non-local" and innovative ceramic classes were retained by Rutter but subdivided on the basis of slipped versus unslipped and enlarged to four classes. His Solidly Painted and Burnished class exhibits a dark-firing slip all over the surface of the vessel, which is then burnished. His other three classes are burnished but according to Rutter remain unslipped. The Fine Burnished Non-Gray and Medium Coarse Burnished classes display dark brown, gray-black to mottled surfaces. The emphasis on black as the surface color of the vessels in these classes implied to Rutter that at the end of the firing process reducing conditions were achieved (1995: 20-25). The Fine Gray Burnished vessels represent perhaps the most distinctive "non-local" class of the Lerna IV assemblage. This ware was unslipped, burnished and reduced with a uniformly gray surface color. It is considered the EH III ancestor of the characteristic MH ware known as "Gray Minyan".

Much like Caskey, Rutter hypothesized that these ceramic classes were introduced to Lerna IV by specialized potters who were part of his proposed "population movement" from central Greece. Specifically, Rutter saw the burnishing of solidly painted and unpainted surfaces and reduction firing as new technological innovations from outside the local environment. Based on the results of the XRD comparative analysis, the authors would like to suggest that the sudden appearance of these ceramic classes in Lerna IV is due in part to the increased use of local illitic clays from the Xerias River system. Likewise, the increased use of these clay sources over time is the result of a locally evolving ceramic technology.

ANALYTICAL TECHNIQUES AND SAMPLE PREPARATION

X-ray diffraction/Reference Samples

Samples used for this analysis were from the Xerias river system: (1) the saprolites (SAP) from the metamorphic window (Figure 3, locations 6 and 7); (2) the yellow and maroon Pindos Claystones (PC) from the upper flysch deposits (Figure 3, location 3); (3) Xerias river clay (XRC) from the upper Argive plain (Figure 4, location 2); and (4) Lerna middle and upper gray sections (LMG/LUG) from the lacustrine sediments of the lower Argive plain (Figure 4, location 1).

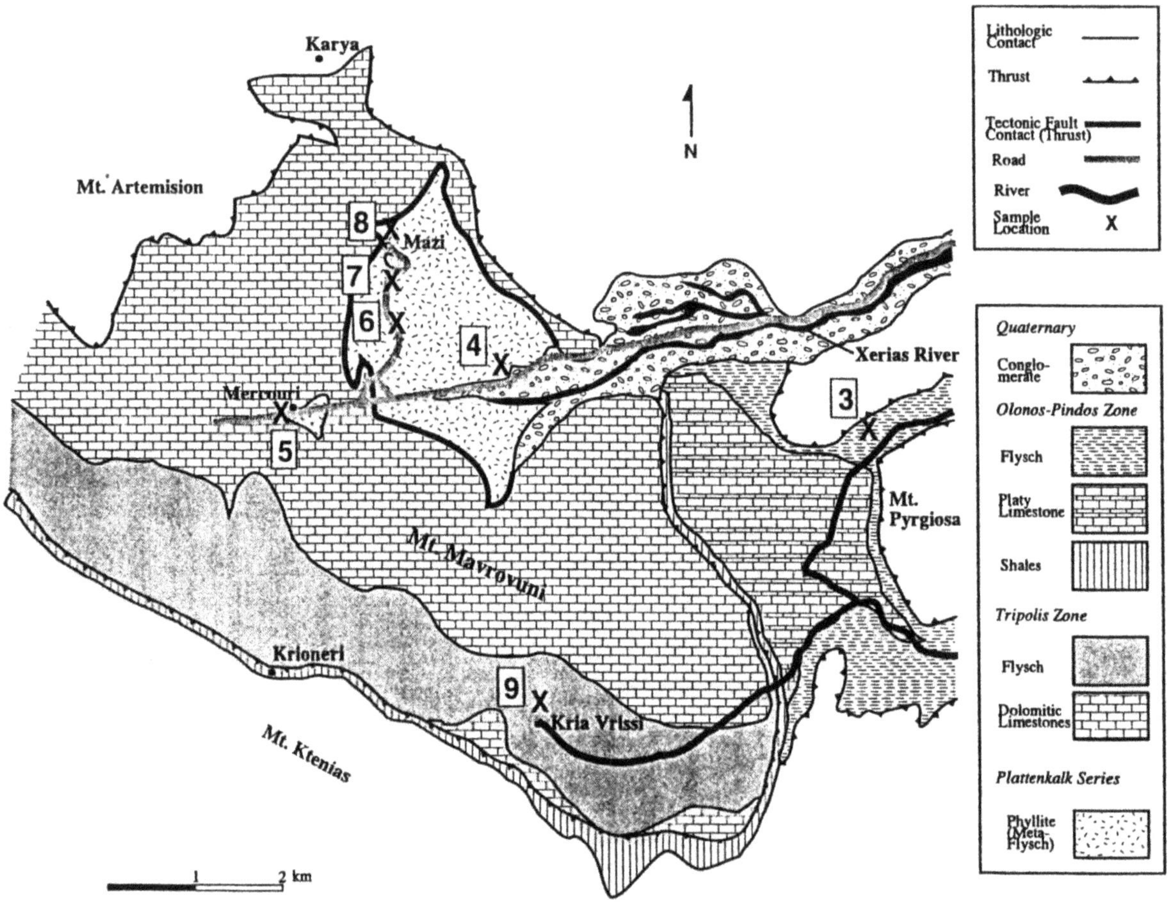

Figure 3. Simplified geological map of the Mazi-Karya metamorphic window, showing the location of the Phyllite-Quartzite Series rocks and the flysches of the Olonos-Pindos and the Tripolis Zone (after Papastamatiou et al. 1970). (X): the location of reference samples.

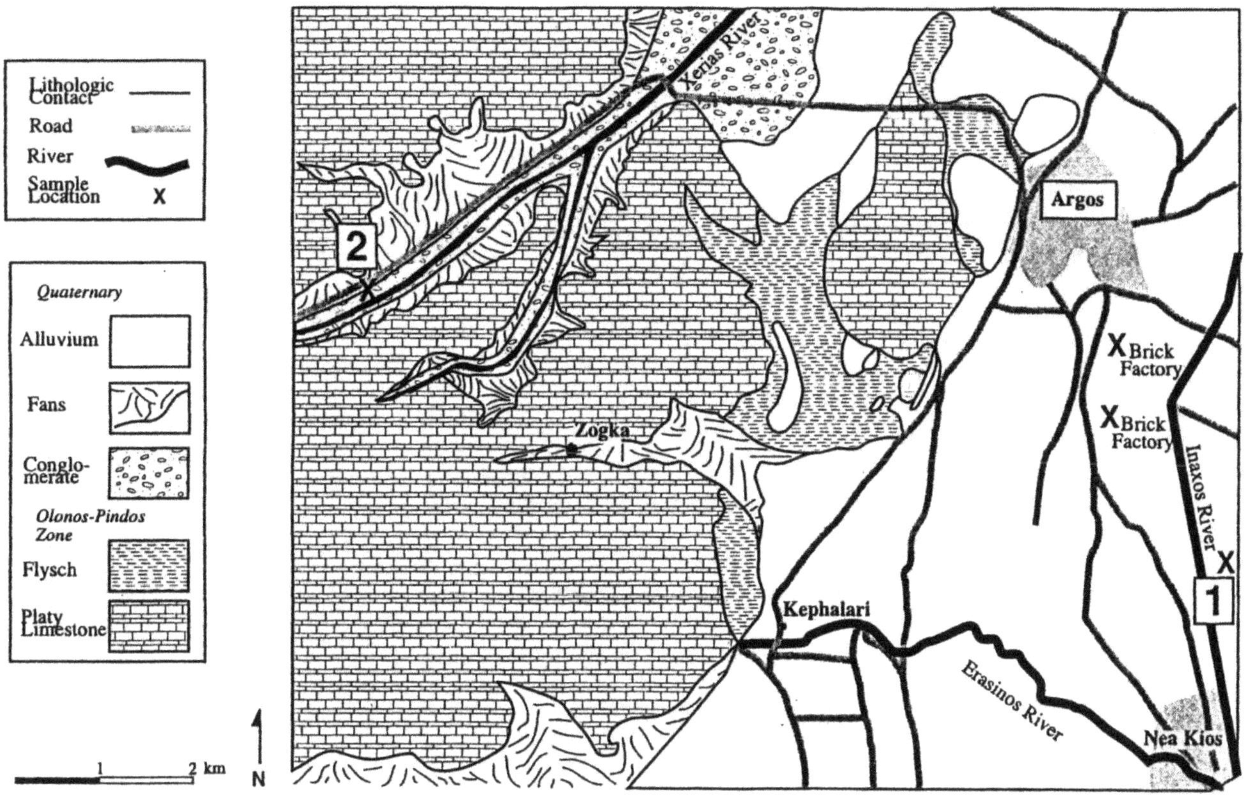

Figure 4: Simplified geological map of the west Argive plain and the Xerias river valley (after Papastamatiou et al. 1970). (X): the location of reference samples.

The raw clays were slaked in de-ionized water, wedged to the point of plasticity and extruded from a Play-Doh brand toy extruder into 1.2x1.2x26 cm bricks. The bricks were fired in a thermal gradient furnace at a heating rate of 10°C/min to 850°C. The bars were segmented into five sections (each spanning ~ 180°) from which one gram samples were collected for micronizing.

One gram samples with a < 0.5 mm particle size were added to the micronizer cup. Ten mls of water were added to the cup, and the samples were ground for 10-15 minutes. The ground samples were washed into 50 ml test tubes and centrifuged to a clear supernate. After decanting the supernates, the ground samples were thoroughly mixed and smeared as thin, glossy coatings on glass slides.

Micronized smear slides were analyzed on a Phillips X-ray diffractometer with Cuk radiation with settings of 45 kilovolts and 20 milliamperes. A scanning rate of 2° 2θ/min over an angular range of 2°- 60° was used. Key minerals and their relative percentages in the reference patterns were used to evaluate specific source locations.

In addition the fired reference clays acted as a comparative tool for determining firing range. Mineralogical and structural transformations occur during the firing process. These changes are identifiable with XRD and depend upon the firing temperature, atmosphere and the mineralogical composition of the raw material. Mineralogical and structural transformations, which occur during the firing process, were used to isolate the firing temperature for the sherd samples. Important mineral transformation parameters utilized in this analysis were (1) calcium carbonate transition to calcium oxide by the evolution of CO_2 at 840-860°C; (2) mullite formation at 930-950°C; (3) kaolinite dehydroxylation at 550°C; and (4) loss of illite structure at 850°C.

X-ray Diffraction Analysis/Artifact Samples

Ceramic sample sets for Lerna III/Phase Late A (Table 1) and Lerna III/Phase Early C (Table 2) were chosen from a 200-count technological sample. Rutter's original petrographic sample set for Lerna IV/Phase 1 was used as the most representative sample for the transition from III to IV (Table 3). Micronized smear slides were prepared for each sherd sample. The artifact patterns from Lerna III/Phase Late A, Lerna III/Phase Early C and Lerna IV/Phase 1 sample sets were compared with the clay reference sample patterns in order to determine source location and firing temperature.

Physical Property Testing/Reference Samples

Green and fired strength along with the percentage of linear firing shrinkage were determined for all the clays which outline the Xerias river system continuum. Modulus of Rupture (MOR) was measured using a Three Point Bend. The reference standard for the MOR tests was a porcelain at 2.440 megapascals (Mpa). Three trials were made for each sample and the results were averaged. Particle size distribution was calculated using a Horiba LA-920, a laser scattering particle size distribution analyzer.

RESULTS

X-ray Diffraction Analysis/Reference Materials

The saprolite samples (SAP1 and SAP2) represent clay-rich sediments collected off of the weathering chloritoid schist with abundant muscovite. The upper section of the schist saprolite (SAP2) is a low-grade kaolinite with associated Fe-rich minerals and the additional non-clay minerals, quartz and feldspar. Kaolinite formation results from the loss of potassium and the reorganization of the alumina and silica due to weathering. A small amount of chlorite is present. SAP1, from one kilometer lower in the profile, is an illite/degraded mica deposit. There is no carbonate in either sample, but a good quantity of the distinctive feldspar is present.

The Pindos claystones are rich in smectite, with some illite and much higher percentages of carbonate than the lower plain alluvial clays. The Pindos claystones are classified as highly calcareous. The Xerias river/upper Argive plain clay (XRC) is predominantly illitic with some smectite and kaolinite. About 5-10% carbonate/calcite is present as a fine carbonate cement. The distinctive feldspar is present along with quartz. This clay is classified as low calcareous as it contains less than 10% Ca.

The lower plain alluvial clays (LMG/LUG) are illitic clays with intergrown smectite layers. The distinctive feldspar is still present, but is being weathered out of the sediment. Thus, the relative abundance of feldspar to carbonate decreases closer to the coastal plain. On the other hand, the amount of calcite present in the alluvial clays increases dramatically. The alluvial clays are classified as calcareous.

Physical Property Testing/Reference Materials

Saprolite 1 and 2 (SAP1&2)
The SAP1&2 clays have a very low green strength of 0.765 - 1.416 Mpa in comparison to the porcelain standard at 2.440 Mpa. Their fired strength, when fired to 900°C, is a range of 1.31 - 3.33 Mpa. Mean particle size is 16.119 - 31.169 µm with a bimodal particle distribution. Shrinkage is -1.5 - 0.25%.

Xerias River Clay (XRC)
The XRC clays have a slightly better green strength of 1.81 Mpa. Their fired strength, when fire to 900°C, is 6.30 Mpa and shrinkage is 0.75%. The mean particle size is 5.441 µm and distribution is bimodal.

Lerna Middle and Upper Gray (LMG/LUG)
The LMG/LUG clays have a high green strength of 2.923 - 3.532 Mpa in comparison to the reference porcelain at 2.440 Mpa. Their fired strength, when fired to 900°C, is a range of 17.16 - 19.73 Mpa. Mean particle size is 4.686 - 5.196 µm with a modal particle size distribution. Shrinkage is 2.25 - 2.75%.

X-ray Diffraction Analysis/Artifacts

All sherd patterns were examined with the comparative reference material pattern superimposed on the sherd pattern. There was no mullite formation in the sherds; however, in many instances the firing temperature was high enough (850°C) that illite was the only remaining clay peak. Recognition of the intensity and shape of the 001 reflection peak for illite and the relative abundance of carbonate to feldspar in the sherd material were important indicators for comparative analysis of source location on the raw material continuum.

Lerna III/Phase Late A
Two clay compositions were noted in the Lerna III/Phase Late A sample (See Table 1). The first clay composition compares to clay sources in the lower Argive plain (LMG/LUG). The second clay composition compares to sources in the upper Argive plain. The upper Argive plain clays are the low calcareous clays (XRC) to metamorphic saprolite (SAP) clays.

The alluvial clay (LMG/LUG) is fired in a very consistent and short firing range of 850-900°C. The SAP/XRC clays display two firing temperatures. The first firing range is between 850-900°C or higher. The second firing range is between 620-775°C. The low-fired samples are notated in Table 1 with an asterisk (*).

Lerna III/Phase Early C
Comparative analysis revealed a similar pattern in the sample for this phase of Lerna III (See Table 2). The alluvial clay (LMG/LUG) is still utilized in Phase Early C and is fired, once again, in the consistently short firing range of 850-900°C. There is a marked increase in the use of the SAP/XRC clays. Within the SAP/XRC technological grouping there is also a more frequent utilization (13.8% in III/A, 24% in III/C) of a lower firing range of 620-775°C. The low-fired samples are notated in Table 2 with an asterisk (*).

Lerna IV/Phase 1
The pattern of two clay compositions, which was observed in the Lerna III phases, is continued in the Lerna IV/Phase 1 sample (See Table 3). The alluvial clay (LMG/LUG) is a far smaller component (65.5% in III/A, 28% in III/C, 34% in IV/1) of the sample than in the Lerna III/Phase Late A sample. The SAP/XRC clays are dominant. Within this technological grouping the low-fired (620-775°C) strategy is relied on to a much greater degree than in the previous phases (13.8% in III/A, 24% in III/C, and 46% in IV/1). The low-fired samples are notated in Table 3 with an asterisk (*).

Summary of the Technological Data

The XRD patterns retrieved from the clay reference samples for the metamorphic saprolite2 (SAP2) and the Xerias River Clay (XRC) confirmed the hypothetical clay matrix progression, observed in the EMPA project, within the metamorphic fabrics for Lerna III and IV. Moreover, comparative XRD analysis between sherd samples and fired reference samples identified two ceramic technological groupings at Lerna (LMG/LUG and SAP/XRC) in the Lerna III and IV sample sets. The LMG/LUG technological grouping can be characterized as a relatively high-fired oxidized ceramic for prehistoric times, which uses a fine to medium fine calcareous clay body. The SAP/XRC technological grouping utilizes both high and lower firing strategies. The high-fired micaceous clay body is a fine to medium coarse fabric; the low-fired is a coarser fabric. The SAP/XRC technological grouping is generally found in conjunction with a burnished surface treatment and/or a reduced surface color. There is an increasing utilization of the low-firing strategy within this grouping in the Lerna III/Phase Early C and Lerna IV/Phase 1 sample sets.

CLAY MINERALOGICAL CONSTRAINTS ON CERAMIC STYLE

The clay mineralogy of the Xerias river system is a function of the local bedrock, which is a series of carbonate platforms overlying a metamorphic window northwest of Lerna. First, there is the poorly sorted quartz phyllite-derived clay weathering in situ on the muscovite-chloritoid schist. Next, the reddish-brown illitic clays of the upper Argive plain are formed through erosional processes and fluvial transport. Finally, the sedimentary illite-smectite clays of the coastal plain are deposited.

Although the top of the metamorphic formation is kaolinitic, the clays of the river continuum are predominantly illitic. There is an increasing calcareous component due to the surrounding carbonate formations. As with most natural samples the illitic clays contain small amounts of interstratified smectite layers. There are three main clay sources along the Xerias river system continuum: (1) the kaolinite from the top of the metamorphic formation northwest of Lerna; (2) the calcareous illitic clays which occur on the highlands and upper Argive plain surrounding Lerna; and (3) the fine-grained calcareous illite-smectite clays of the lower Argive plain. As will be seen below, the calcareous illitic clays offer the best possibilities for the development of a durable, all-purpose and still aesthetically pleasing ware.

LMG/LUG Technological Grouping

The XRD patterns for the LMG/LUG technological grouping are striking in their visual regularity (Figure 5). These clays contain very fine-grained carbonate (calcite) which acts as a flux allowing for a denser, glassy phase at a lower temperature. When fired in the 850-1050°C range, fine-grained calcite occurring in the clay matrix reacts with silica and alumina to produce calcium aluminum silicates such as gehlenite and anthorite; some of the calcium may be incorporated into a glassy phase (Heimann 1989: 136).

The black core associated with these clays and observed in the fired thermal gradient bricks is not an indication of low-firing; but is the black coring associated with rapid firing which leads to vitrification. Black coring indicates a temperature gradient from outside, which is high enough to burn out the carbon, whereas there is still carbonaceous material in the core, which indicates a lower internal temperature.

Early PE assessment of Lerna III thin-sections suggested that they represented a bisque firing (900-1000°C). The carbonate dust evenly distributed across the matrix is naturally occurring and an indication of this type of firing. A bisque fired body can reduce porosity to 2-3% (Rice 1987: 106). This characteristic would be valuable in shapes, which imply use as a fine or tableware, e.g. sauceboat, saucer, pyxis, and cup. A drawback for the use of this clay could be the relatively high firing temperature and narrow firing range that must be attained and controlled in order for the fine particle calcite to sinter and fuse (Murray 1994: 247; Rice 1987: 104).

The considerable green strength of the LMG/LUG clays allows for significant structural integrity. This physical characteristic is particularly important in the construction of large ceramic pieces where sagging of the body needs to be avoided, e.g. pithoi (storage jars). The presence of the small amount of smectite in these clays increases plasticity and dry strength. The 2.25 - 2.75% shrinkage is not high enough to cause structural flaws (Murray 1994: 247).

This technological grouping does have a small range in particle size. Clays with bimodal or trimodal particle size distribution usually lead to better particle packing/density. A wide particle size distribution will help to lower porosity and increase fired strength. A small distribution can cause higher porosity and weaker fired strength, except when smectite is present in small quantities.

SAP/XRC Technological Grouping

This clay group exhibits from Lerna III/Phase Late A onwards the capability for two firing strategies. The fine to medium coarse clay (SAP) is fired to the same range as the LMG/LUG grouping but is not as hard and glassy as the LMG/LUG clays due to the lack of a fluxing agent (Figure 6).

The second firing strategy, which encompasses calcareous illitic clays (XRC), places much more emphasis on a low-fired, porous and coarser-textured ceramic. The carbonate can be employed as a flux, which lowers firing temperature, reduces fuel needs and still achieves a dense ceramic body (Figure 7).

A coarser, thicker-walled and lower-fired ware is better suited for cooking where thermal shock resistance is important. Firing strategies, however, are needed in order to control firing temperature and reduce the effects of spalling. The thermal decomposition of calcite to CaO in the ceramic and the increase in volume caused by recarbonation after cooling can lead to destruction of the pot. If the calcareous illitic clay is fired under reducing rather than oxidizing conditions, the decomposition reactions of calcite and illite are coupled. No free CaO occurs and pots have less risk of being destroyed by lime spalling (Heimann 1989:136).

The results of physical property testing suggest that the SAP clays lack the green and fired strength of the LMG/LUG technological grouping. The saprolites do have a bimodal size distribution, which should help fired strength and lower porosity. Their large particle size and absence of small quantities of smectite makes them less plastic. The XRC clays of this technological grouping exhibit a higher green and fired strength as well as bimodal particle size distribution. The bimodal particle distribution is in part a reflection of the natural blending of the micaceous and calcareous components of the Xerias river system continuum. Shrinkage is much lower than for the LMG/LUG grouping.

DISCUSSION

The raw material continuum that was observed through the EMPA study revealed in the compositional analyses of the clay matrices for Lerna III and IV sherds the suggestion that between these two cultural periods there was a change in specific source locations, perhaps reflecting changes in preference for particular purposes. The original Lerna III and IV petrography reports of Betancourt and Myer (2000: 680-703) and Betancourt et al. (1995: 666-693) show, as well, a marked increase in the utilization of micaceous clays (SAP/XRC) by the later phases of Lerna III/C and D and into Lerna IV/Phase 1.

The XRD analysis confirmed the probe evidence and the fabric patterning that is apparent in Betancourt and Myer's Lerna III and IV petrography. In Lerna III/Phase Late A the Lernean potters used for their metamorphic fabrics a more muscovite-rich clay derived from higher in the river system. The coarser, more carbonate-rich illitic clay source has a small presence. Both the fine and coarse metamorphic fabrics are completely overshadowed by the utilization of fine-grained alluvial clays (LMG/LUG).

By Lerna III/Phase Early C the Lernean potters were consistently utilizing two sources for their metamorphic fabrics: (1) a finer-grained micaceous clay; and (2) a coarser more carbonate-rich illitic clay from lower in the Xerias river system. The XRD sample set shows the same increase in preference for these two clay sources that was seen in the original petrography sample. The consistent utilization of the two sources from the upper Xerias river system continued in the Lerna IV/Phase 1 samples. Indeed, the large, specialized and thin-walled Lerna III assemblage of bowls, saucers and storage vessels with few handles gives way slowly to a smaller, less-varied Lerna IV assemblage of coarser and thicker-walled vessels with many more handles present (Rutter 1993).

A majority of Rutter's presumed "non-local" and innovative classes come from the Disparate Metamorphic PE group (1995: 736-749). This is the same group that we have been able to identify as local and have designated as the SAP/XRC technological grouping. Thus, we infer that the use of burnished surface treatments and reduction firings which are observed in the Disparate Metamorphic PE fabric group (and the ceramic classes represented by that PE fabric) throughout both Lerna III and IV can be discussed in terms of local development. How much of the technological innovation exhibited in the Lerna IV classes of Fine Gray Burnished, Fine Burnished Non-Gray, Medium Coarse Burnished and Solidly Painted and Burnished could be the result of a locally evolving ceramic technology?

There is a tendency towards partially reduced vessels in both LMG/LUG and SAP/XRC technological groupings, but reduction firing is primarily associated with the SAP/XRC clays. Reduction firing becomes much more apparent within the SAP/XRC technological grouping in the Lerna IV/Phase 1 assemblage. A reduction atmosphere during the firing process will harden the finished product. Under reducing conditions iron compounds react with silica to produce a flux. This reaction creates a glassy phase at a lower temperature (Rice 1987: 354). Philpots has shown that the burning of grass temper during firing can be used to produce reducing conditions. His conclusion was that "by mixing organic temper with the clay the potter insured that reducing conditions were achieved. The reduced ware allowed for a less permeable vessel at lower firing temperatures" (1994: 616). Wiencke has mentioned an organic/vegetal temper, which appears in the later phases of Lerna III (2000: 318).

In the sample sets burnishing or polishing as a surface treatment is most prevalent in the SAP/XRC technological grouping. Rutter notes that "a larger percentage of those vases coated all over with paint and subsequently burnished have a fine, rather soft, and slightly micaceous fabric" (1995: 54). Slipped and burnished surfaces retard the penetration of liquids by creating a dense surface of fine and compacted particles (Rice 1987: 231). The use of this surface treatment is often a response to a less sintered ceramic body. Although there is a greater fondness in Lerna IV for burnished vessel surfaces, burnishing associated with painted and unpainted surfaces is documented by Wiencke for Lerna III as well (2000: 320-328).

Based on the evidence from the sample sets, there is an increasing frequency of pots produced in SAP/XRC clays. Is there a technological explanation for this trend toward a coarser clay with more carbonate and/or quartz present in the body? The comparative firing temperature analysis revealed that the Limestone with Metaquartzite PE fabric group, i.e. the XRC clays, in the Lerna III/Phase Late A, Lerna III/Phase Early C and Lerna IV/Phase 1 sample sets is virtually always a low-fired ceramic. One suggestion might be that the Lernean potters wanted the properties of strength and resistance to thermal shock in specific ceramic shapes fired at the lower firing temperature. To accomplish this they needed a clay body with a fluxing agent to create the necessary phase reactions for an inert body at a lower temperature. Heimann has shown that calcareous illitic clays were widely used in antiquity, because they have a lower vitrification temperature than kaolinite-rich clays (1989: 132).

The greater use over time of coarser clays fired at a lower firing temperature could suggest a change in the importance of specific local clay sources. More coarse wares in Lerna IV could imply more on-site cooking and food preparation – and the possibility of more permanent households at Lerna in IV than in III. One could speculate that in EH II (Lerna III) Lerna was a trading post and a habitation site in Lerna IV.

Another suggestion might be that these technological groupings of upper Xerias river clays and lower valley alluvial clays represent two populations of potters. The highland population had at its disposal two clay compositions; the lower plain population had only the alluvial clays. Each population could have eventually specialized in products for which their clays were best suited. If the technological groupings do represent two local populations at Lerna, the changes in clay preferences and the increasing frequency of the SAP/XRC technological grouping in the sample sets could imply a significant change in local settlement regimes at Lerna.

Shriner has made one observation, which might tend to discount, but not preclude, the idea of two separate populations of potters. At the same time (Lerna III/C and D) that the Lernean potters were utilizing an upper Xerias river clay more frequently for fine and special purpose ceramics, for their house tiles they were moving away from the fine micaceous clay of the SAP/XRC technological grouping to a calcareous clay. Prior to the "House of Tiles" (Lerna III/Phase D), roofing tiles for special buildings were produced in a micaceous clay. In contrast, the "House of Tiles" is roofed in tiles produced from a calcareous clay (LMG/LUG technological grouping). It appears that the Lernean potters were aware of both upper Xerias river and lower valley alluvial clays from an early period. With time they learned the characteristics and advantages of each kind of clay body and came to use their raw materials for the specific purposes most suited to the clay body. The less porous tiles executed in the high-fired alluvial clays would be far superior to tiles constructed from micaceous clays. Indeed, these alluvial clays are still used in modern times for the construction of roof tiles.

ARCHAEOLOGICAL IMPLICATIONS

The technological evidence suggests the continuous utilization of two clay sources and their dependent technologies throughout cultural periods III and IV at Lerna. The SAP/XRC clays represent the local metamorphic window northwest of Lerna and the upper Xerias river system; the LMG/LUG clays represent the lower Argive plain. Beginning in Lerna III/Phase Early C, the SAP/XRC clays appear to have increased in popularity. Coupled with this more diverse use of local clays in Lerna III/Phases C and D, an increasing technological sophistication in firing strategies is documented. There appears to be a trend toward using a clay body that is chemically stable/vitrifies at lower firing temperatures. The ceramic technology data sets support a continuity in ceramic technological practices among local potters between the Lerna III and IV ceramic assemblages.

The technological evidence suggests that non-local influences on local Lernean populations and potters at the Lerna III/IV boundary were of a nature that implies equal status between foreign elements and indigenous elements, i.e. trade, small groups of newcomers and the diffusion of ideas. What may appear as a sudden stylistic ceramic change between Lerna III and IV could be interpreted as a series of much earlier and more gradual adaptations by local potters to a changing social and economic environment. These social and economic factors could include: (1) imported forms or immigrant craftsmen; (2) new forms and refurbished old forms adapted to accommodate new technologies and changes in social behavior; (3) the introduction of new foods that necessitated changes in storage or cooking vessels; (4) the implications of marriage; (5) specific individuals or groups who influenced the use of specific styles; and (6) technological utilization of local clay compositions for specific purposes or physical characteristics.

The ceramic stylistic evidence for the transition between Lerna III and IV at Lerna documents a discontinuity in shape evolution, decorative painting and surface color. At the same time the technological evidence suggests a defensible continuity among the ceramic classes and the long-term use of local clay compositions within local potter groups. The combined stylistic and technological evidence from Lerna suggests an overprint – not replacement – on a local ceramic assemblage. In the history of Lerna ceramics this overprinting on an existing assemblage may signal a significant and persistent local population of artisans.

Acknowledgments
The authors would like to express their gratitude to Dr. William Carty, Director of the Whiteware Research Center, New York State College of Ceramics at Alfred University, for conducting the physical property testing on the Lerna reference clays. Special thanks are extended to Ms. Erin Becker, who performed the tests and organized the results. Funding for fieldwork and sample development was provided for C.S. by the Fulbright Foundation, the Harriet and Leon Pomerance Fellowship, the Program in Classical Archaeology (Indiana University) and the Wiener Laboratory (American School of Classical Studies at Athens).

References

Betancourt, P.P. and Myer, G.H. 2000. Petrographic Examination of Pottery Samples from Lerna III, In *Lerna, A Preclassical Site in the Argolid, IV, Lerna III: Architecture, Stratification and Pottery*, M.H. Wiencke. Princeton, NJ: American School of Classical Studies at Athens, pp. 680-703.

Betancourt, P.P., Myer, G.H., and Vaughan, S.J. 1995. Ceramic Petrography of Selected Pottery Samples from Lerna IV. In *Lerna, A Preclassical Site in the Argolid, III, The Pottery of Lerna IV*, J.B. Rutter. Princeton, NJ: American School of Classical Studies at Athens, pp. 666-693.

Caskey, J. L. 1960. The Early Helladic Period in the Argolid. *Hesperia* 29: 285-303.

Deer, W.A., Howie, R.A., and Zussman, J. 1992. *An Introduction to the Rock-Forming Minerals*. 2nd edition. Hong Kong: Longman Scientific and Technical.

Heimann, R.B. 1989. Assessing the Technology of Ancient Pottery: The Use of Ceramic Phase Diagrams. *Archeomaterials* 3: 123-148.

Jones, R.E. 1986. *Greek and Cypriot Pottery: A Review of Scientific Studies*. Fitch Laboratory Occasional Paper 1. Athens: British School at Athens

Jones, R.E. 1993. Pottery as Evidence for Trade and Colonisation in the Aegean Bronze Age: the Contribution of Scientific Techniques. In *Wace and Blegen: Pottery as Evidence for Trade in the Aegean Bronze Age, 1939-1989*, eds. C. and P. Zerner and J. Winder, Amsterdam: Gieben, pp. 11-17.

Matson, F.R. 1965. Ceramic Ecology: an Approach to the study of the Early Cultures of the Near East. In *Ceramics and Man*, ed. F.R. Matson. Viking Fund Publications in Anthropology, 41. Chicago: Wenner Gren Foundation, pp. 202-217.

Murray, H.H. 1994. Common Clay. In *Industrial Minerals and Rocks*, 6th edition, ed. D.D. Carr, Littleton, Co: Society for Mining, Metallurgy and Exploration, Inc., pp. 247-248.

Papastamatiou, J., Vetoulis, D., Tataris, A., Christodoulou, G., Bornova, J., Lalechos N., and Kounis, G. 1970. *Geological Map of Greece. 1:50,000*. Argos Sheet. Athens: ISGR.

Philpots, A. 1994. Application of Petrofabric and Phase Equilibria Analysis to the Study of a Potsherd. *Journal of Archaeological Science* 21: 607-618.

Rice, P. 1987. *Pottery Analysis: A Sourcebook*. Chicago: University of Chicago Press.

Rutter, J.B. 1993. Early Helladic Pottery: Inferences about Exchange and Production from Style and Clay Composition. In *Wace and Blegen: Pottery as Evidence for Trade in the Aegean Bronze Age, 1939-1989*, eds. C. and P. Zerner and J. Winder, Amsterdam: Gieben, pp. 19-37.

Rutter, J.B. 1995. *Lerna, A Preclassical Site in the Argolid, III, The Pottery of Lerna IV*. Princeton, NJ: American School of Classical Studies at Athens.

Shriner, C. and Dorais, M.J. 1999. A Comparative Electron Microprobe Study of Lerna III and Lerna IV Ceramics and Local Clay-Rich Sediments. *Archaeometry* 41(1): 25-49.

Vaughan, S.J., Myer, G.H., and Betancourt, P.P. 1995. Discussion and Interpretation of the Petrographic Data. In Lerna, *A Preclassical Site in the Argolid, III, The Pottery of Lerna IV*, J.B. Rutter. Princeton, NJ: American School of Classical Studies at Athens, pp. 693-710.

Weaver, C.E. 1989. *Clays, Muds, and Shales*. Amsterdam: Elsevier Scientific.

Whitbread, I.K. 1995. *Greek Transport Amphorae: a Petrological and Archaeological Study*, Fitch Laboratory Occasional Paper 4, British School at Athens, Oxford.

Wiencke, M.H. 2000. *Lerna, A Preclassical Site in the Argolid, IV, Lerna III: Architecture, Stratification and Pottery*. Princeton, NJ: American School of Classical Studies at Athens.

Table 1. Technological sample for Lerna III/Phase A. The sample was used for X-ray diffraction analysis. Low-fired (660-775 °C) samples are notated with an asterisk (*). Source: Wiencke 2000.

LERNA III/PHASE LATE A					
Techno Fabric	**PE Fabric**	**Shape**	**Surface Treatment**	**Macrofabric**	**Sample**
LMG/LUG	Fine w/Carbonate	Askos	Dark Painted, Patterned	Medium Fine	P218
		Jug or Jar	Dark Painted, Patterned	Medium Fine	P125
		Pyxis	Dark Painted, Patterned	Medium Fine	P169
		Sauceboat	Dark Painted, Patterned	Medium Fine	P173
		Saucer	Dark Painted, Patterned	Medium Fine	P128
		Basin	Dark Painted	Medium Coarse	P223
		Basin	Dark Painted	Medium Fine	P133
		Jar	Dark Painted	Medium Fine	P136
		Jar	Dark Painted	Medium Fine	P227
		Jar	Dark Painted	Medium Coarse	P225
		Jug	Dark Painted	Medium Fine	P137
		Pyxis	Dark Painted	Medium Fine	P141
		Sauceboat	Dark Painted	Medium Fine	P143
		Sauceboat	Dark Painted	Medium Fine	P142
		Saucer (Pedestal)	Dark Painted	Medium Fine	P145
		Saucer	Dark Painted	Medium Fine	P148
		Saucer	Dark Painted	Medium Fine	P152
		Jar	Unpainted	Medium Fine	P161
		Jug	Unpainted	Medium Fine	P162
SAP/XRC	Fine/ Micaceous	Jug	Light Painted, Fine Polished	Fine	P117
		Bowl (Pedestal)	Light Painted, Burnished	Medium Coarse	P123
		Saucer	Light Painted, Burnished	Medium Coarse	P124
		Saucer	Unpainted	Medium Coarse	P120
	Metamorphic	Basin	Light Painted, Fine Polished	Medium Coarse	P122
		Askos	Unpainted	Medium Coarse	P159
	Limestone w/Metaquartzite	Basin	Light Painted, Fine Polished	Medium Fine w/Mica	P116*
		Bowl (Incurved)	Coarse	Very Coarse	P164*
		Pithos	Coarse	Coarse	P211*
		Hearth	Coarse	Coarse	P210*

Table 2. Technological sample for Lerna III/Phase Early C. The sample was used for X-ray diffraction analysis. Low-fired (660-775 °C) samples are notated with an asterisk (*). Source: Wiencke 2000.

LERNA III/PHASE EARLY C					
Techno Fabric	**PE Fabric**	**Shape**	**Surface Treatment**	**Macrofabric**	**Sample**
LMG/ LUG	Fine w/Carbonate	Jar or Jug	Dark Painted, Patterned	Medium Fine	P579
		Basin	Dark Painted	Medium Coarse	P547
		Sauceboat	Dark Painted	Medium Coarse	P548
		Saucer	Dark Painted	Medium Coarse	P556
		Basin	Unpainted	Medium Coarse	P558
		Pyxis	Light Painted, Fine Polished	Very Fine	P607
		Sauceboat	Light Painted, Fine Polished	Fine	P543/T569
SAP/XRC	Fine/Micaceous	Saucer	Light Painted, Fine Polished	Medium Fine	P608
		Bowl	Dark Painted	Medium Coarse	P563
		Bowl (Pedestal)	Light Painted	Medium Coarse	P575
		Basin	Light Painted, Burnished	Medium Coarse	P544
		Saucer	Light Painted, Fine Polished	Fine	P570
		Saucer	Light Painted, Fine Polished	Fine	P571
		Jar (Hydria)	Dark Painted	Medium Coarse	P628
	Metamorphic	Jar	Light Painted, Burnished	Medium Coarse	P545/T568
		Jar (Pithos-Sized)	Unpainted	Medium Coarse	P560
		Baking Pan	Coarse	Coarse	P605
SAP/XRC	Limestone w/Metaquartzite	Basin	Dark Painted	Medium Coarse	P627*
		Bowl (Collared)	Light Painted	Medium Fine	P574*
		Pithos (Bowl T.)	Dark Painted, Coarse	Coarse	P562
		Pithos (Bowl T.)	Coarse	Coarse	P565*
		Pithos (Necked T.)	Coarse	Coarse	P566
		Jar	Unpainted	Medium Coarse	P559*
		Hearth	Unpainted	Coarse	P690*
	Siltstone	Bowl (Ring-Base)	Coarse	Coarse	P564*

Table 3. Original ceramic petrography sample for Lerna IV/Phase 1. The sample was used for X-ray diffraction analysis. Low-fired samples are notated with an asterisk (*). Sources: Rutter 1995; Betancourt et al. 1995.

LERNA IV/PHASE 1

Techno Fabric	PE Fabric	Shape	Surface Treatment	Macrofabric	Sample
LMG/LUG	Fine w/ carbonate	Jar (Narrow-Necked)	Dark on Light, Painted	Very Fine	P119/T514
		Jar (Narrow-Necked)	Dark on Light, Painted		P120/T516
		Jar (Narrow-Necked)	Dark on Light, Painted		P126/T518
		Tankard	Dark on Light, Painted		P20/T517
		Tankard	Dark on Light, Painted	Fine	P386/T519
		Tankard	Dark on Light, Painted	Very Fine	P2/T513
		Ouzo Cup	Light on Dark, Painted	Fine	P378/T533
		Tankard	Solidly Painted and Burnished	Fine/Medium Coarse	P168/T368
		Bowl	Solidly Painted and Unburnished		P197/T379
		Ouzo Cup	Solidly Painted and Unburnished		P179/T378
		Tankard	Fine Gray Burnished	Fine	P201
		Kantharos	Fine Gray Burnished	Fine	P202
SAP/XRC	Fine/ Micaceous	Jug	Light on Dark, Painted	Fine	P286/T531
		Cup/Bowl	Light on Dark, Painted	Highly Micaceous	P287/T532
		Bowl	Solidly Painted and Burnished	Very Fine w/Mica	P331/T369*
		Tankard	Solidly Painted and Burnished	Fine w/Mica	P164/T367*
		Bowl	Fine Burnished Non-Gray	Fine	P210/T358
		Bowl	Fine Burnished Non-Gray	Fine	P213/T359
		Tankard	Fine Burnished Non-Gray	Fine w/Mica	P206/T350
		Lid	Fine, Unburnished	Very Fine w/Mica	P301/T364*
		Tankard	Medium Coarse, Burnished	Medium Coarse	P320/T387
		Open Shape	Coarse	Coarse	P256/T405
SAP/XRC	Limestone w/ Metaquartzite	Askos	Fine Burnished Non-Gray	Fine w/Mica	P216/T357*
		Bowl	Fine Burnished Non-Gray	Fine	P296/T360*
		Bowl	Fine Burnished Non-Gray	Fine	P219/T361*
		Bowl	Fine Burnished Non-Gray	Fine	P292/T388*
		Saucer	Fine, Unburnished	Fine w/Mica	P310/T363*
		Cup	Coarse	Coarse	P241/T538*
		Jar (Wide-Mouthed)	Coarse	Coarse	P352/T408*
		Jar (Wide-Mouthed)	Coarse	Coarse	P330/T537*
		Tankard	Coarse	Coarse w/Mica	P408/T543*
	Siltstone	Cup	Medium Coarse, Burnished	Medium Coarse	P313/T386*
		Cup	Medium Coarse, Burnished	Medium Coarse	P321/T406*
		Cup	Medium Coarse, Unburnished	Medium Coarse	P322/T400*
		Jug	Dark on Light, Painted	Medium Coarse	P111/T515*

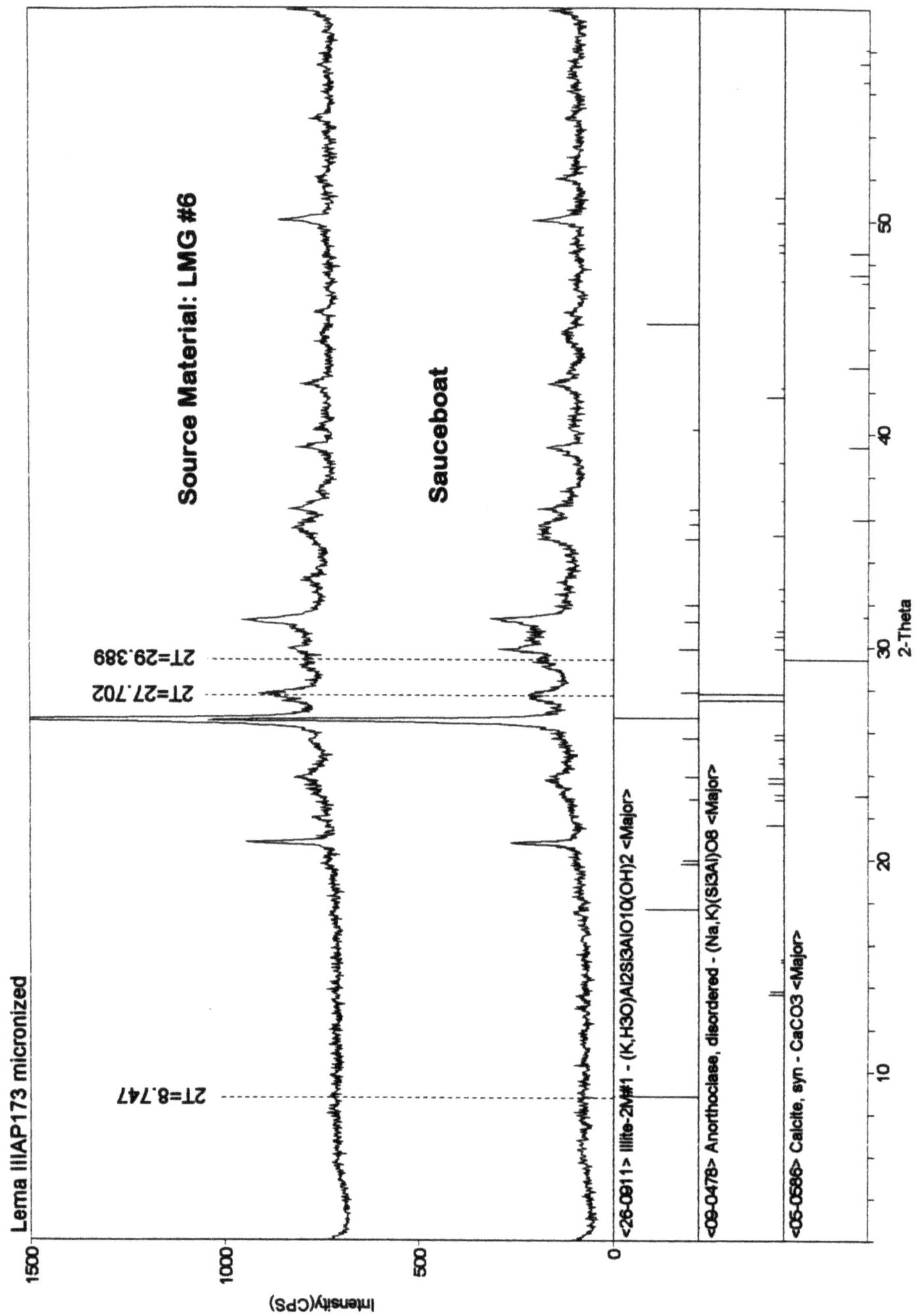

Figure 5. Comparative XRD pattern for Lerna III/A P173.

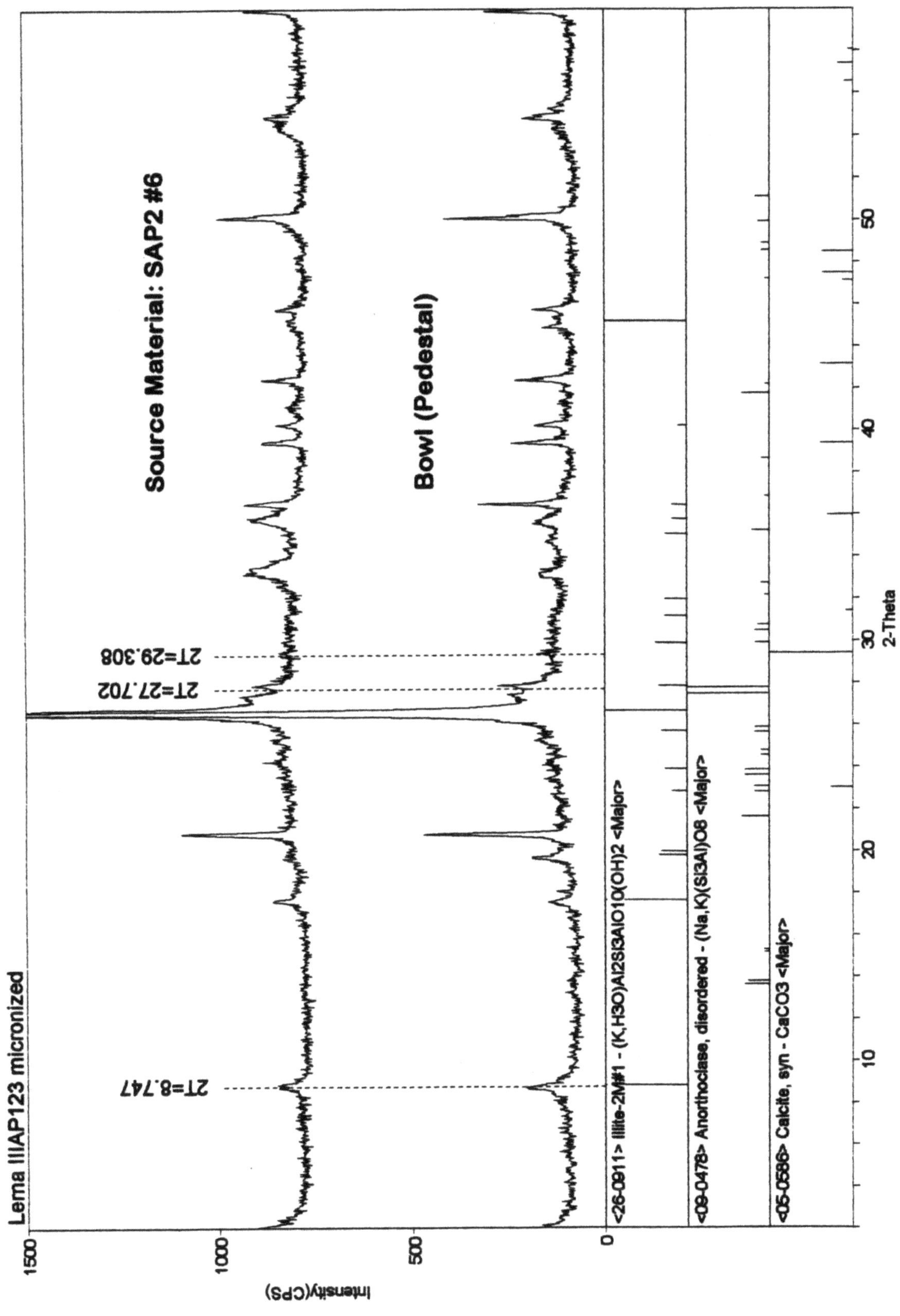

Figure 6. Comparative XRD pattern for Lerna III/A P123.

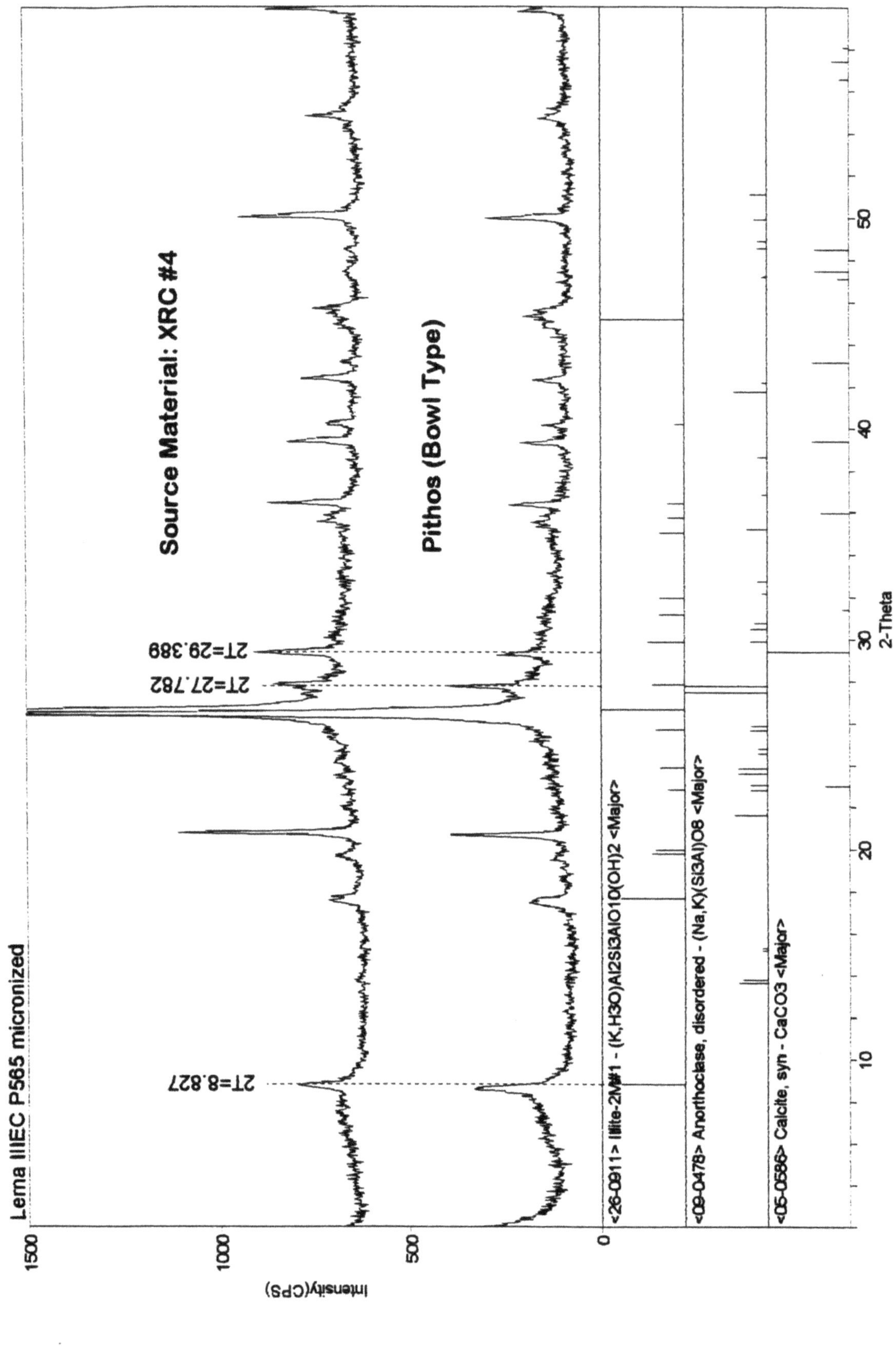

Figure 7. Comparative XRD pattern for Lerna III/EC P565.

CHAPTER 2

ANATOLIAN METALLIC WARE: A THIRD MILLENNIUM BC CERAMIC PHENOMENON

Elizabeth S. Friedman
Department of Near Eastern Languages and Civilizations, University of Chicago

Five thousand years ago, ancient Near Eastern potters created "metallic wares", hard, thin, ceramics identified by the metallic sound they make when tapped or struck. Based on their sound and appearance, metallic wares have been touted as examples of early, high-fired ceramics. Anatolian metallic ware, in particular, has been described as high fired, despite the absence of any scientific evidence supporting such a claim. In order to investigate this metallic phenomenon, samples of Anatolian metallic ware were examined by multiple analytical techniques to determine the degree of vitrification, the elemental composition, and mineral component of the clay. Phase analysis by scanning electron microscopy reveals that the Anatolian metallic ware exhibits the physical qualities of a high-fired ceramic, but without vitrification. Compositional analysis by energy dispersive spectrometry and neutron activation analysis establishes that the metallic ware from Anatolia was manufactured from a phyllosilicate, serpentine clay. The use of this high magnesium - low potassium clay source enabled the potters of ancient Anatolia to produce a brittle, dense, non-porous ceramic without being fired at high temperatures.

INTRODUCTION

During the third millennium BC or Early Bronze Age, a new type of pottery called "metallic ware" appears in ceramic assemblages throughout the Near East. This pottery is not dubbed "metallic ware' because it contains large quantities of metal, but rather because of the metallic ring it makes when tapped or struck against a hard surface. Metallic ware represents a technological innovation that stretches from central Anatolia to the southern Levant. Contemporary Early Bronze Age metallic wares have been found in south and central Turkey (Goldman 1956; Mellaart 1954, 1963; Mellink 1989, 1993; Özten 1989), the 'Amuq Valley (Braidwood and Braidwood 1960), northern Israel (Greenberg and Porat 1996), and northeastern Syria (Kühne and Schneider 1988; Schneider 1988, 1989). Each of these regions exhibits a metallic ware ceramic tradition differentiated by the composition of the clay, choice of temper, mode of manufacture, range of vessels, surface treatment, firing temperature, and firing atmosphere (for a detailed comparison of Near Eastern metallic wares see Friedman 2000). For decades, noted archaeologists and pottery specialists such as Robert and Linda Braidwood (1960), James Mellaart (1962), Machteld Mellink (1993), Frederick Matson (1956), Amnon Ben-Tor (1992), Rafael Greenberg and Naomi Porat (1996) have all assumed that the metallic ring of metallic ware was an effect of high-firing temperatures. Research was undertaken to test this assumption and more broadly, to explain the relationship between metallic wares and metal production (see Friedman 2000). Of particular interest to this study is the Anatolian metallic ware from south-central Turkey, where copper and bronze production proliferated during the third millennium BC.

The Anatolian metallic ware sherds selected for analysis come from the small, Early Bronze Age site of Göltepe, located in the Taurus Mountains, about 80 kilometers north of the Mediterranean coast. Archaeological excavations at Göltepe revealed subterranean structures, pottery, and evidence of metal processing primarily in the form of clay crucibles, casting molds, and grindstone tools (Vandiver, Yener, and May 1992; Yener and Vandiver 1993). The neighboring tin mine of Kestel appears to have been exploited at about the same time. Ceramic evidence combined with radiocarbon dates place the occupation of Göltepe and the mining of Kestel to between ca. 2800-2200 BC.

Although Göltepe is only 8-10 hectares in size, contemporary cultural debris in the form of ceramic material and lithics covers about 65 hectares (Yener and Vandiver 1993). The industrial ceramics are represented by thousands of fragments of crude, hand-made chaff-tempered crucibles. All of the domestic pottery was handmade as well. The household assemblage includes orange-gritty pithoi, low-walled baking trays, hole-mouth cooking pots, cups, shallow vessels, and deep bowls. The bowls and cups were manufactured in coarse ware, thick red-slipped ware, micaceous ware (both burnished and unburnished), grog-tempered fine-slipped ware, and Anatolian metallic ware. The jugs and pitchers are restricted to Anatolian metallic ware.

ANATOLIAN METALLIC WARE

Third millennium BC Anatolian metallic ware is found primarily in south-central Turkey (Figure 1). Its distribution ranges from the central Konya plain, east to the foothills of the Taurus Mountains, and from Acemhöyük on the central plateau, south to the Cilician coast (Garstang 1953; Garstang and Goldman 1947; Goldman 1956; Mellaart 1954, 1963; Mellink 1993; Özgüç 1986; Özten 1989; Seton-Williams 1954; Yener and Vandiver 1993). Sherds of Anatolian metallic ware have been found in and around the Cilician Gates in the Taurus Mountains and along the Calycadnus pass at the eastern edge of the Konya plain, two major overland routes providing access to the Cilician Plain (Özten 1989).

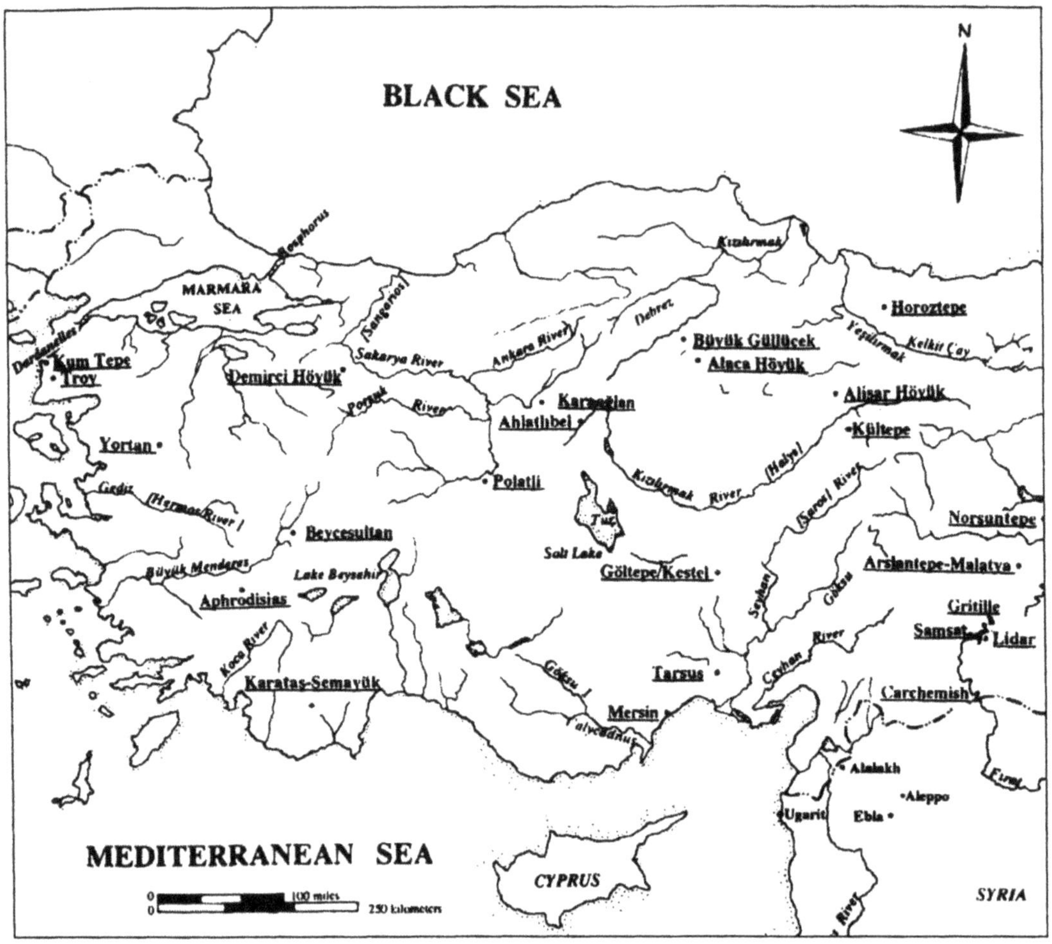

Figure 1. Map of selected Early Bronze Age sites in central Turkey (adapted from Joukowsky 1996, fig. 5.1, p. 142).

A preliminary look at the chronological and geographical distribution of Anatolian metallic ware suggests that the technology probably originated in the central and eastern half of the Konya Plain sometime during the first quarter of the third millennium BC (EB I). By the mid-third millennium BC (EB II), Anatolian metallic ware, which was still prominent on the Konya plain, spread eastward toward the foothills of the Taurus Mountains and into Cilicia. At Göltepe, Anatolian metallic ware firmly dates to the EB II and continues into the latter half of the third millennium BC (EB III). In the EB III, Anatolian metallic ware appears at Acemhöyük north of the Taurus range (Özgüç 1986; Özten 1989), but by that time, it has all but vanished from the Konya and Cilician Plains (Friedman 2000).

Anatolian metallic ware is a hard, handmade, thin-walled (2-5mm) ceramic that breaks along a straight fracture line. The clay used for producing the metallic ware is well levigated to create fine, durable, non-porous vessels suitable for storage, transport, and serving liquids. There are no traces of organic temper nor is there evidence for the deliberate addition of non-plastic inclusions. The repertoire of Anatolian metallic ware consists primarily of globular jars and jugs (Figure 2). The jars have slight to moderate rising spouts with a single vertical handle attached to the rim and shoulder. The jugs have straight or slightly everted collars with one or two vertical handles attached to the neck and shoulder, although there are handleless variants as well. The shapes include bowls, single-handled cups, and cups with tubular spouts. All of the vessels have slightly concave or omphalos bases. Lug handles, attached both vertically and horizontally, are often found on the vessels with handles. Although Özten (1989) suggests that the lug handles are decorative, it is more likely that they are functional, probably used for tying on cloth or leather lids.

To date, more than thirty different incised or impressed patterns have been found on the handles from Göltepe. The incisions consist of combinations of dots and dashes. Similar handles with incisions or pot marks have been found on other sites in the Niğde region, Ereğli, and Acemhöyük (Mellaart 1954, 1963; Özten 1989). These marks denote the vessel's contents, the intended recipient, the potter's workshop, or may simply reflect a range of decorative designs.

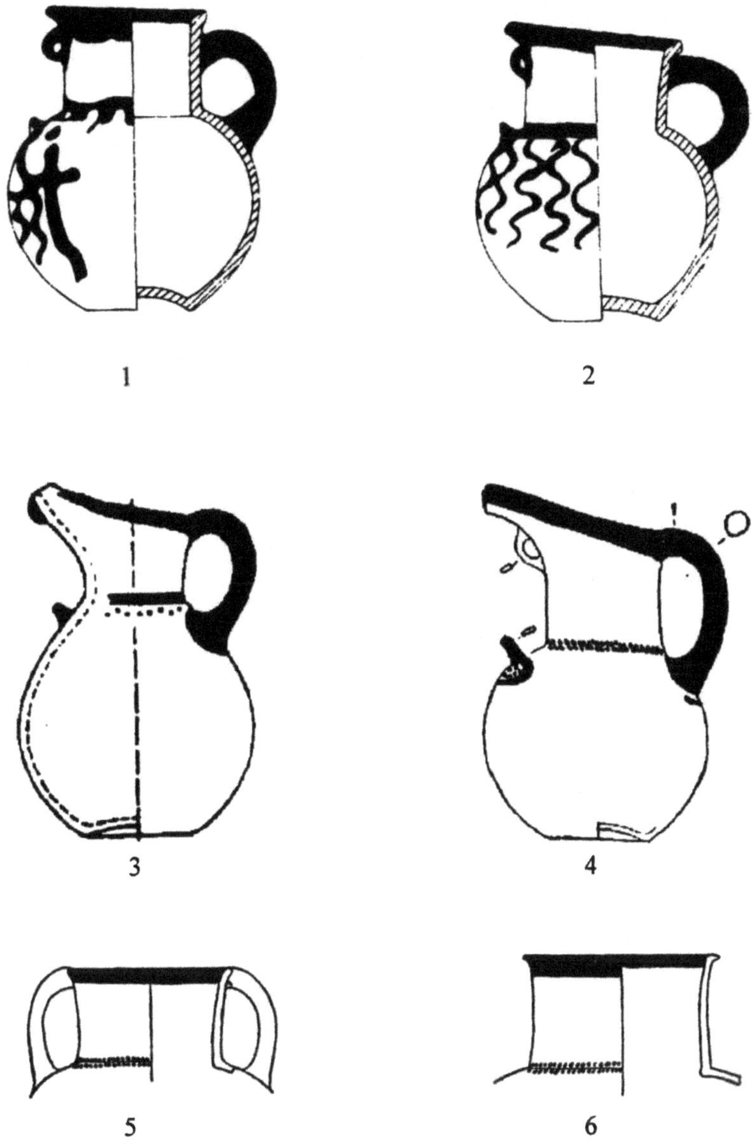

Figure 2. Examples of Anatolian metallic war. Vessels 1 and 2 from Özten (1989: 416); vessel 3 from Garstang (1953: 196); and vessels 4, 5, and 6 from Mellaart (1963: 231,233). Vessels not drawn to scale.

The interior surface of the Anatolian metallic-ware sherds collected from Göltepe is fairly consistent, registering reddish yellow (7.5 YR 7/6) on the Munsell chart. Variants include light gray (10 YR 7/2), very pale brown (10 YR 7/4), pink (7.5 YR 7/4), and very dark gray (5 YR 3/1). The core is generally reddish yellow (7.5 YR 6/6) but varies from light yellowish brown (2.5 Y 6/3) to pink (7.5 YR 7/4). The exterior is generally reddish yellow (5 YR 7/6) but has many more variants than the interior, including light brown (7.5 YR 6/3-6/4), pink (7.5 YR 7/4), reddish brown (5 YR 5/3), brown (7.5 YR 4/3) and even dark gray (2.5 Y 4/1). It sometimes even appears with a greenish tinge. The inconsistency in the exterior color of the sherds suggests that the firing conditions were not completely uniform. Özten notes (1989) that a thin slip is applied either in the color of the clays itself or in one or two shades darker, and that a wash is generally present, though this is not easily detected. High, moderate, and even slight burnish is readily visible.

Anatolian metallic ware is often decorated with a purplish-red or purplish-brown design. The most common pattern is a wide, simple band painted around the rim, both inside the vessel and out. A single row of dots, a line of squiggles, or rows of dots are often found encircling the base of the neck, separating the neck from the shoulder. Wavy vertical lines occasionally appear on the upper half of the body. Mellaart (1963) notes that the paint is thin and washy and Özten (1989) adds that the dripping effect was a result of paint applied with a thin brush before firing. An occasional tree-like motif can be found on the neck or on the shoulder near the handle.

THE ARCHAEOLOGICAL ASSUMPTION

The archaeological assumption that the sound of metallic ware, its appearance, and its brittleness, is a result of high-firing temperatures applies to all Early Bronze Age metallic wares. In a discussion of "thin sandy ware", a variant of Anatolian metallic ware from Tarsus, Matson remarks that the hardness and metallic ring, together with its gray core suggests that the pottery was fired to a fairly high temperature in a reducing atmosphere (Matson 1956: 360). Following Matson, Mellink (1993: 500) suggests that the manufacture of Cilician red gritty ware and pottery with metallic characteristics required that the vessels be "overfired," which resulted in a brittle, clinky fabric. Mellaart (1962: 21) commented that one of the classes of pottery that marked the onset of the EB I on the Konya Plain was gritty stoneware [metallic ware] that was fired to a point where the clay became vitrified. Similarly, Yener and Vandiver suggest that the clinky characteristic of Anatolian metallic ware is due to it being "hard fired" (1993: 217). All of the metallic-related wares from south-central Anatolia, including sandy ware, red gritty ware, and Anatolian metallic ware, have been associated with high-firing temperatures.

Similarly, the metallic wares from Israel and northern Syria have been said to be high fired (Ben-Tor 1992; Greenberg and Porat 1996; Schneider 1988, 1989). In a discussion on EB III Canaanite pottery, Ben-Tor states that the term "metallic" applies to vessels that have been exceptionally well fired. The higher firing temperature produces a harder, stronger, and more durable pot, which, when struck, gives off a high-pitched metallic sound (Ben-Tor 1992: 108). Greenberg and Porat (1996: 5) also define metallic ware by its highly fired fabric. As for the few recovered sherds of metallic ware from the 'Amuq, the Braidwoods comment that the pottery has a "fine musical ring" and that the "core is rather vitrified," while Matson adds that the clay was high fired (Braidwood and Braidwood 1960: 370). All of the remarks regarding hard, thin, durable metallic-like ceramics are based upon visual, audible, and tactile attributes.

Despite the apparent consensus that ancient Near Eastern metallic ware was high fired, no materials science report has ever been published supporting such a conclusion for the Anatolian variety. Therefore, it seemed appropriate to analyze this ceramic material to determine whether or not it was actually high fired, to see if it showed any signs of vitrification. Multiple analytical techniques were used to obtain the elemental and mineral composition of the fabric as well as the phase of vitrification. The ceramic analysis incorporated scanning electron microscopy, energy dispersive spectrometry, neutron activation analysis, optical microscopy, and x-ray diffraction.

PHASE ANALYSIS USING SCANNING ELECTRON MICROSCOPY

Scanning electron microscopy (SEM) is an excellent technique for determining the relative degree to which pottery has been fired (Maniatis and Tite 1978, 1981; Tite and Maniatis 1975; Tite, Freestone et al. 1982; Tite, Maniatis et al. 1982). The SEM produces backscattered electron images or photomicrographs that reveal the phases of vitrification. As the temperature rises, the clay particles melt and subsequently fuse, thereby creating a glassy matrix containing gas-filled spherical pores. The continuity and extent of the glass phase increases as the temperature rises. The scanning electron images clearly display the degree to which the clay particles have created such a glassy matrix.

Vitrification starts with sintering, the process of bonding between clay particles. Sintering is activated after the water molecules have evaporated, at temperatures between 500-800°C, depending upon the type of clay, the firing temperature, firing atmosphere, and duration of firing. At about 800-850°C the body mass starts shrinking and the interstitial pores are either eliminated or filled with gases. As the clay shrinks, the vessel's density increases. With the increase in density, porosity decreases, and the vessel becomes stronger. Consequently, the characteristics of density and strength are directly related to the degree to which the ceramic is vitrified. Hence, wares that are fired at higher temperatures should be more durable than wares fired at lower temperatures. Therefore, one would expect Anatolian metallic ware to be high fired, since it possesses the physical properties of low-porosity and vessel strength.

Scanning electron microscopy in a low-vacuum mode was used for backscattered imaging of four polished thin-sections of Anatolian metallic ware, two of which are presented here. The samples were analyzed with a JEOL JSM-5800LV SEM equipped with a Charge Free Anticontamination System, which precluded the necessity for coating. The backscattered images do not show any indications of a glassy matrix. The photomicrographs of the Anatolian metallic ware do not reveal the phases of vitrification that were anticipated from the hard, thin-walled ceramic. Figures 3 and 4 demonstrate that the clay platelets have indeed begun to fuse but that the temperature or length of firing was not sufficient for the clay to undergo full vitrification. The lack of vitrification implies the absence of high firing and strongly suggests that the pottery was fired at temperatures below 1000°C. After examining the backscattered images, Mike Tite determined that the Anatolian metallic ware was probably fired between 800-900°C (1997: personal communication). Thus, the metallic ware from Göltepe appears to have been fired at temperatures lower than those used for the metallic Syrian stoneware, which was fired between 1000-1100°C (Schneider 1988), and below the minimum firing temperatures required for smelting tin and copper. Since the Anatolian metallic ware was clearly not high fired, it was obvious that other chemical or physical properties must contribute toward its thin, hard, durable character.

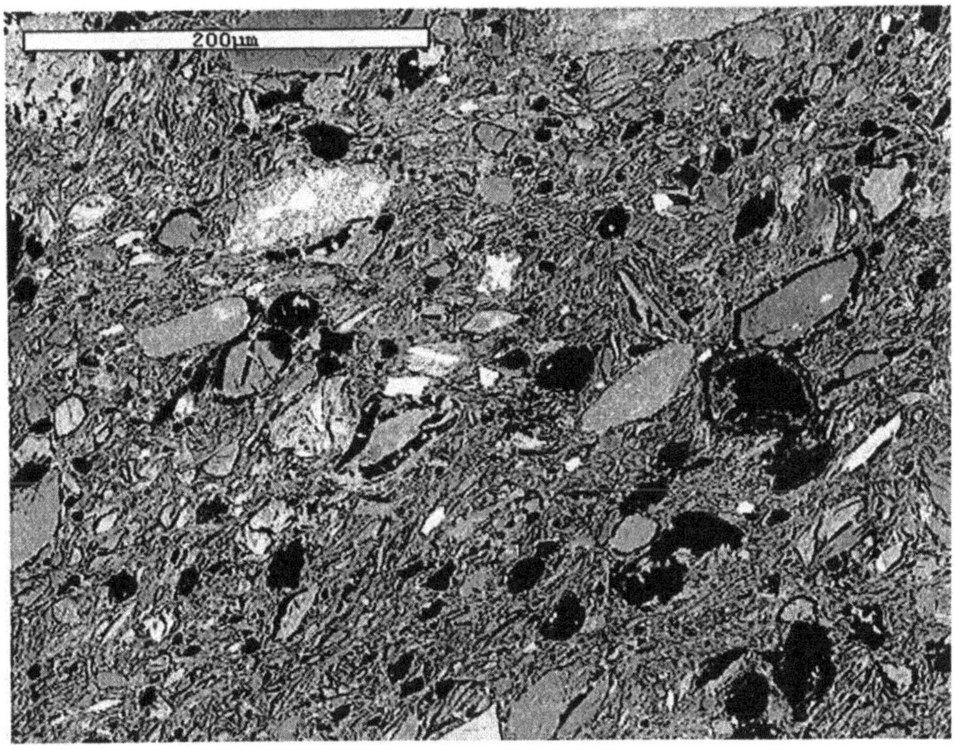

Figure 3. Backscattered electron image of Anatolian metallic ware from Göltepe. The clay platelets have not yet undergone vitrification.

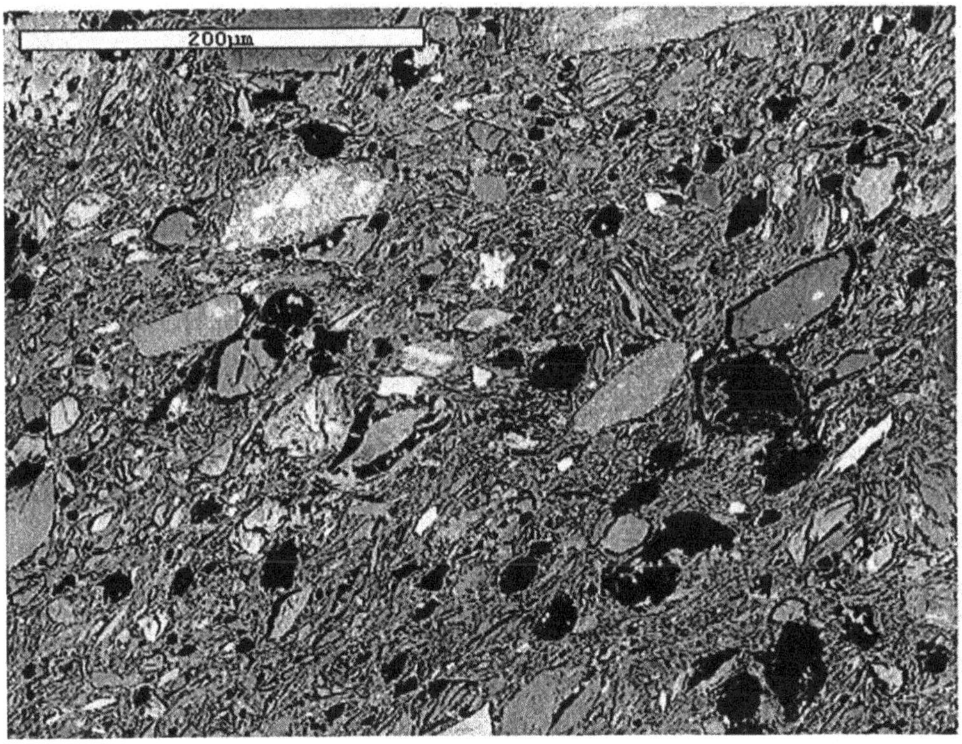

Figure 4. Backscattered electron image of Anatolian metallic ware from Göltepe. There are no signs of a glassy matrix or gas-filled spherical pores.

ELEMENTAL COMPOSITION USING ENERGY DISPERSIVE SPECTROMETRY

To obtain the chemical composition of the metallic ware, polished, thin, cross-sections were examined by energy dispersive analysis using an ISIS-LINK system fitted to the scanning electron microscope. The points for x-ray analysis were hand-selected so that the chemical composition of the clay could be obtained exclusive of the mineral inclusions, which consisted primarily of quartz, feldspars, and mica. The energy dispersive spectrometry measured the major elements in oxide form: Na_2O, MgO, Al_2O_3, SiO_2, K_2O, CaO and FeO. Twenty points were measured for each of four thin-sections, the data of two of which are presented herewith. Internal standards were provided by the LINK system.

Energy dispersive analysis was conducted on three groups of ceramics from Göltepe: Anatolian metallic ware; burnished, micaceous, and crucible wares; and grog-tempered fine-slipped ware. It should be noted that the burnished and micaceous wares were typed in the field by the difference in their appearance. The micaceous ware was full of bright flecs of mica whereas the burnished ware was smooth and glossy. Upon elemental analysis, it was discovered that the burnished and micaceous wares were manufactured from the same mixture of clay and temper. Burnished ware is simply micaceous ware that has undergone surface treatment. Therefore, from the materials science perspective, the two wares should be treated as one. The crucibles are grouped with the burnished and micaceous wares based on the similarity of their elemental compositions. The third group, fine-slipped ware, is chemically and visually distinct from the other wares.

The oxide compositions acquired by the energy dispersive analysis of three groups of ceramics from Göltepe are presented in Tables 1, 2 and 3. The data in Table 1 reveal that Anatolian metallic ware was manufactured from a magnesium-silicate clay containing on average 28% magnesium oxide (MgO). This contrasts sharply with the data obtained from the group of burnished, micaceous, and crucible wares, which contain only 2.5-4.5% MgO (Table 2). The fine-slipped ware contains even less MgO, only about 1.5% (Table 3). The Anatolian metallic ware also differs from the two other groups by its alumina (Al_2O3) content. Anatolian metallic ware contains on average 7-8.5% Al_2O3 whereas the burnished, micaceous, crucible, and fine-slipped wares contain 16-21% Al_2O3. This high amount of alumina is characteristic of aluminum-silicate clays. In addition to the amounts of magnesium oxide and aluminum oxide, the selected wares from Göltepe also differ in the amount of potassium oxide or potash (K_2O). Anatolian metallic ware contains less than 0.5% K_2O while the burnished, micaceous, crucible, and fine-slipped wares contain on average 2-4% K_2O.

It can be inferred from these data that the low amount of potash may have contributed to the hard-fired qualities of the Anatolian metallic ware. Potassium generally behaves like a flux, lowering the temperature at which vitrification takes place. Without potassium, the clay platelets are more likely to vitrify only when the ceramics are fired at higher temperatures. Thus, the absence of vitrification may be due to the potassium deficient nature of the clay. But what role did the magnesium play? The high magnesium - low potassium composition of the Anatolian metallic ware pottery was intriguing. The association of magnesium with magnesium-rich minerals had to be determined in order to figure out the clay source used by these ancient potters.

To identify the mineral compound, a polished thin-section of Anatolian metallic ware was placed in a polarizing light microscope. Under polarized light, the magnesium particles appeared black, revealing a nearly isotropic or non-crystalline structure. At first this suggested that the "minerals" were actually particles of glass. However, this was highly improbable considering that glass was not part of the material culture of Early Bronze Age Anatolia. Therefore, the most likely candidate for the isotropic particles would be crushed obsidian, an abundant and prominent natural resource in south-central Turkey. Yet, none of the obsidian sources or obsidian artifacts from the ancient Near East contains greater than 0.5% magnesium, with most averaging far less than that (Cauvin et al. 1999; Gratuze 1999; Keller and Seifreid 1990). Having ruled out obsidian and glass, other methods were used to help determine the nature of the isotropic magnesium particles.

Table 1. Göltepe Anatolian metallic ware clay matrix

Compound %

Sample	Na_2O	MgO	Al_2O3	SiO_2	K_2O	CaO	TiO_2	FeO
MW-2336-g	1.77	28.20	7.48	48.78	0.42	6.34	0.36	6.65
MW-2336-I	1.73	27.05	6.96	49.07	0.45	7.63	0.24	6.88
MW-3722-j	1.69	29.33	8.54	48.86	0.10	3.95	0.18	7.36
MW-3722-k	1.64	27.96	7.49	51.37	0.17	3.87	0.19	7.30

Table 2. Burnished, micaceous, and crucible clay matrix

Compound % Sample	Na_2O	MgO	Al_2O_3	SiO_2	K_2O	CaO	TiO_2	FeO
BB-4888-a	0.75	4.01	18.44	53.98	2.07	12.89	1.16	6.71
BB-4888-b	0.84	3.54	18.43	58.34	2.36	9.91	0.69	5.89
BB-4919-a	1.29	4.39	14.52	59.44	3.15	11.43	0.29	5.47
BB-4919-b	1.55	4.34	19.34	56.62	1.97	8.89	1.32	5.97
MIC-2336-e	1.03	3.81	19.93	56.26	2.47	9.73	0.47	6.29
MIC-2336-f	0.86	4.74	18.22	59.81	2.72	7.15	0.65	5.85
MIC-4381-a	2.29	2.58	20.34	62.19	3.43	3.97	0.40	4.80
MIC-4381-b	1.69	4.06	19.66	59.80	3.00	5.04	0.76	5.99
CRU-4549b-3	4.03	2.38	19.89	58.78	3.91	6.40	0.73	4.18
CRU-4549b-6	2.28	2.76	20.78	60.05	3.38	3.16	1.41	6.18
CRU-4591b-5	4.80	3.03	20.28	62.06	3.53	1.00	0.36	4.91
CRU-4591b-6	4.98	1.69	19.18	63.95	5.97	1.14	0.26	2.71

Table 3. Fine-slipped ware clay matrix

Compound % Sample	Na_2O	MgO	Al_2O_3	SiO_2	K_2O	CaO	TiO_2	FeO
FS-4141-m1	1.41	1.71	16.63	64.91	1.83	8.37	0.40	4.75
FS-4141-m2	1.63	1.61	20.91	61.53	2.16	6.81	0.33	5.03
FS-4372-mm	3.22	1.34	18.53	65.60	3.28	4.56	0.25	3.21
FS-4372-mn	3.14	1.52	16.62	68.08	2.53	4.47	0.33	3.30

In addition to the optical microscopy, a small amount of ceramic material from an Anatolian metallic ware potsherd was selected for x-ray diffraction analysis. This sample was representative of the type of clay from which the ceramic was made but because of the cryptocrystalline structure of the mineral, the results were inconclusive. The multiple analytical techniques employed thus far revealed a non-vitrified ceramic material manufactured from a high-magnesium clay source composed of materials with a non-crystalline structure. Because x-ray diffraction could not assist in the determination of this mineral, further analysis of Anatolian metallic ware was required. Although the energy dispersive data provided the major elemental composition, in order to obtain a more complete chemical fingerprint, it was necessary to supplement this information by trace element analysis.

TRACE ELEMENT ANALYSIS USING NEUTRON ACTIVATION ANALYSIS

Although energy dispersive spectrometry (EDS) has the ability to measure concentrations of naturally occurring major elements in clay, it does not have the sensitivity to detect trace elements. Neutron activation analysis (NAA), on the other hand, while not capable of measuring some of the major elements, has the ability to perform bulk analysis of 30 or so elements, including trace elements. In the case of Anatolian metallic ware, it was crucial that magnesium was detected by EDS, since magnesium could not be detected by NAA and magnesium oxide turned out to be a strong portion of the chemical fingerprint. At the same time, the trace element data provided by NAA, in particular the measurements of chromium and nickel were just as important to the establishment of the chemical fingerprint.

Neutron activation analysis was performed on 11 samples of Anatolian metallic ware from Göltepe. The metallic ware sherds were obtained primarily from occupation surfaces, room fill, and trash pits, all within the subterranean structures or pit houses at Göltepe. In addition, four other classes of ceramic material from the Göltepe assemblage were analyzed along with two local sources of alluvial clay. The ceramic assemblage included 13 samples of crucibles, 9 samples of burnished ware, 5 samples of micaceous ware, and 12 samples of fine-slipped ware. The primary objectives of the neutron activation analysis were to determine the number of "compositional groups" and to establish the geochemical relationship between these groups. We were also interested in determining the probability that some of the ceramic chemical fingerprints matched the fingerprint of the local clays, thus raising the possibility of local manufacture (De Sena and Friedman 1997; Friedman 2000).

Neutron activation results for the ceramics from Göltepe demonstrate that Anatolian metallic ware is a distinct, cohesive group and that the probability that the samples of metallic ware are related to any other group from the Göltepe ceramic assemblage is extremely low. Nor does the composition of Anatolian metallic ware match the local alluvial clays, which have a close geochemical relationship to the crucibles, micaceous, and burnished wares (for details see De Sena and Friedman 1997; and Friedman 2000). The concentrations of trace elements are quite variable, particularly for elements such as chromium and nickel, elements that occur in much greater amounts in the Anatolian metallic ware than in the other wares (Table 4). The high values for chromium and nickel along with the high magnesium oxide - low potassium oxide composition clearly contribute to the chemical fingerprint of Anatolian metallic ware.

Table 4. Concentrations of chromium and nickel – Group mean (ppm)

Ceramic Groups	No. of Samples	Chromium	(s.d.)	Nickel	(s.d.)
Anatolian Metallic Ware	(11)	1870.00 ppm	(7.8%)	908.00 ppm	(6.3%)
Fine-Slipped Ware	(12)	52.90 ppm	(23.6%)	24.80 ppm	(41.5%)
Crucibles and Burnished Ware	(13)	166.00 ppm	(36.2%)	60.50 ppm	(42.6%)
Micaceous and Burnished Ware, Crucibles, and Clay Samples	(13)	341.00 ppm	(32.8%)	116.00 ppm	(45.6%)

CLAY MINERALOGY

Neutron activation analysis of eleven specimens of Anatolian metallic ware measured an average of 1870 ppm of chromium and 908 ppm of nickel. These two elements, chromium and nickel, are often found together in serpentinite, which is derived from ultrabasic and igneous rocks. Serpentinite is named after the mineral serpentine [$Mg_3Si_2O_5(OH)_4$], a hydrated magnesium silicate. Serpentine is a group of three primary minerals that have the same chemical composition but vary in their physical and optical properties: antigorite, lizardite, and chrysotile (asbestos). Of the three minerals, the most likely candidate for Anatolian metallic ware is antigorite. When antigorite is fired the structure is destroyed. This dehydroxylated material is amorphous to x-rays, thus explaining why the fired material could not be identified by x-ray diffraction or polarizing light microscopy.

The fact that antigorite is a phyllosilicate mineral is important for understanding the potter's choice of serpentine clay. Antigorite, with its greasy luster, is massive, fine-grained, and difficult to fuse (Klein and Hurlbut 1985; Dana and Dana 1997). The clay does not readily absorb water and does not easily expand. The inability to accommodate water in the structure of the clay suggests that this type of clay would be suitable for manufacturing vessels of a non-porous nature. Upon drying and firing, shrinkage is insignificant and porosity is reduced. The low porosity enables the clay particles to bind tightly together. The fabric, therefore, becomes dense without necessarily being thick. The high density and low porosity enable the ceramic to achieve a high stage of hardness without having to undergo vitrification. It is perhaps this density that gives the ceramic its metallic-sounding ring (Shepard 1985, 131).

Burnishing the magnesium-silicate based ceramic also contributes to its hardness and impermeability. To achieve the effect of burnishing, a polishing stone is rubbed across the vessel before it is fired. The force of the hard object on the vessel wall aligns and depresses the clay particles, thereby reducing the spaces between them. The compression of the particles increases the density, consequently strengthening the vessel wall while at the same time reducing the porosity. Since light is reflected by the alignment of the clay particles, the greater the force of the burnish, the greater the shine. Appearance not withstanding, the potters created a hard, dense, impermeable ceramic material. Due to the potter's discriminating choice of clay sources, Anatolian metallic ware possesses all of the qualities of high-fired ware without being high fired.

CONCLUSIONS

Metallic wares first appear in the ancient Near East in archaeological contexts dating to the third millennium BC. The thin, hard, non-porous fabrics are readily identified by the sound the potsherds produce when tapped or struck. This clinky or metallic sound has been attributed to high firing. Anatolian metallic ware in particular has been heralded as a fine archaeological example of a high-fired ceramic. Scanning electron microscopy, energy dispersive spectrometry, neutron activation analysis, optical microscopy, and x-ray diffraction show that the physical characteristics of Anatolian metallic ware are the result of specific chemical and mineralogical attributes, not vitrification. The high magnesium, chromium, and nickel suggest the presence of antigorite, which is derived from serpentinite. The deliberate selection of a magnesium-silicate or serpentine clay with a very low potassium content enabled the potters to produce a ceramic that did not have to undergo vitrification to obtain its strength, durability, and impermeable nature.

Each and every one of the analytical techniques has its limitations, but when used together they provide a wealth of information about ancient ceramic material. Not only do the analyses reveal that Anatolian metallic ware was not necessarily fired at high temperatures, but equally significant is the finding that Anatolian metallic ware was manufactured from a type of clay that had not been previously exploited for ceramics. Although serpentine is a common natural resource in south-central Anatolia, it was not used for pottery until the Early Bronze Age. How the potters discovered that they could use serpentine-derived clay to produce pottery is not known and why it ceased to be exploited for manufacturing ceramics at the end of the Early Bronze Age requires further investigation.

Acknowledgments

I would like to thank Aslihan Yener, Behin Aksoy, Fazil Açikgöz, and Erol Faydali, in assisting me in acquiring the potsherds and granting me permission to conduct the analyses. The SEM/EDS analysis was conducted by the author at the Henry Hinds Laboratory at the University of Chicago under the patient supervision of Dr. Andrew Davis, senior scientist with the Enrico Fermi Institute. Many thanks to Dr. Ian Steele of the Department of Geophysical Sciences, University of Chicago for the optical microscopy and x-ray diffraction analysis. The neutron activation analysis was performed at the nuclear reactor at the University of Illinois at Urbana-Champaign (UIUC) in collaboration with Eric de

Sena, with the assistance of Dr. Sheldon Landsberger, and with the encouragement of Dr. Sarah Wisseman. The financial resources for the reactor costs were provided by the DOE reactor-share program. Travel expenses were generously provided by a Ryerson Fellowship.

References

Alp, S. 1968. *Zylinder- und Stempelsiegel aus Karahöyük bei Konya.* Türk Tarih Kurumu yayinlarindan, series 5, no. 26, Ankara.

Ben-Tor, A. 1992. The Early Bronze Age. In *The Archaeology of Ancient Israel*, ed. A. Ben-Tor. New Haven: Yale University Press, pp. 81-125.

Braidwood, R.J., and Braidwood, L.S. 1960. *Excavations in the Plain of Antioch I: The Earlier Assemblages: Phases A-J.* Oriental Institute Publications vol. 61. Chicago: University of Chicago Press.

Cauvin, M.C., Balkan, N. Besnus, Y., and Saroglu, F. 1999. Origine de L'obsidienne de Cafer Höyük (Turquie): Premiers Résultats. *Paléorient* 12/2 (1986): 89-97.

Dana, J.D., and Dana, E.S. 1997. *Dana's New Mineralogy*, 8th ed. New York: John Wiley & Sons.

De Sena, E.C., and Friedman, E.S. 1997. An Investigation of the Sources of Bronze Age Pottery from Göltepe and Kestel, Turkey by Neutron Activation. In *Materials Research Society Symposium Proceedings: Material Issues in Art and Archaeology* 5, vol. 462, ed. P.B. Vandiver et al. Pittsburgh: Materials Research Society, pp. 239-246.

Friedman, E.S. 2000. *Technological Style in Early Bronze Age Anatolia: The Interrelationship between Ceramic and Metal Production at Göltepe.* Ph.D. dissertation. Chicago: Department of Near Eastern Languages and Civilizations, University of Chicago.

Garstang, J. 1953. *Prehistoric Mersin: Yümük Tepe in Southern Turkey, The Neilson Expedition in Cilicia.* Oxford: The Clarendon Press.

Garstang, J., and Goldman, H. 1947. A Conspectus of Early Cilician Pottery. *American Journal of Archaeology* 51: 370-388.

Goldman, H. 1956. *Excavations at Gözlü Kule, Tarsus II: From the Neolithic through the Bronze Age.* Princeton: Princeton University Press.

Gratuze, B. 1999. Obsidian Characterization by Laser Ablation ICP-MS and Its Application to Prehistoric Trade in the Mediterranean and the Near East: Sources and Distribution of Obsidian within the Aegean and Anatolia. *Journal of Archaeological Science* 26: 869-881.

Greenberg, R., and Porat, N. 1996. A Third Millennium Levantine Pottery Production Center: Typology, Petrography, and Provenance of the Metallic Ware of Northern Israel and Adjacent Regions. *Bulletin of the American Schools of Oriental Research* 301: 5-24.

Joukowsky, M.S. 1996. *Early Turkey: Anatolian Archaeology from Prehistory through the Lydian Period.* Dubuque, Iowa: Kendall/Hunt Publishing.

Keller J., and Seifried, C. 1990. The Present Status of Obsidian Source Identification in Anatolia and the Near East. *Journal of the European Study Group on Physical, Chemical, Biological and Mathematical Techniques Applied to Archaeology* (PACT) 25 (4): 57-87.

Klein, C., and Hurlbut Jr., C.S. 1985. *Manual of Mineralogy* (after J.D. Dana), 20th ed. New York: John Wiley & Sons.

Kühne, H., and Schneider, G. 1988. Neue Untersuchungen zur Metallischen Ware. *Damaszener Mitteilungen* 3: 83-139.

Maniatis, Y., and Tite, M.S. 1978. Ceramic Technology in the Aegean World during the Bronze Age. In *Thera and the Aegean World*, vol. 1., International Scientific Congress on the Volcano of Thera (2) 1978: Santorini Greece. London: Thera and the Aegean World, pp. 483-492.

Maniatis, Y., and Tite, M.S. 1981. Technological Examination of Neolithic-Bronze Age Pottery from Central and Southeast Europe and from the Near East. *Journal of Archaeological Science* 8: 59-76.

Matson, F.R. 1956. Appendix: Techniques of the Early Bronze Potters at Tarsus. In *Excavations at Gözlü Kule, Tarsus II: From the Neolithic through the Bronze Age*, ed. H. Goldman. Princeton: Princeton University Press, pp. 352-36.

Mellaart, J. 1954. Preclassical Remains in Southern Turkey. *Anatolian Studies* 4: 175-239.

Mellaart, J. 1962. Anatolia c. 4000-2300 B.C. In *The Cambridge Ancient History*, vol. 1, Chapter 18. Cambridge: Cambridge University Press, pp. 3-53.

Mellaart, J. 1963. Early Cultures of the South Anatolian Plateau, II. *Anatolian Studies* 13: 199-236.

Mellink, M. 1989. Anatolian and Foreign Relations of Tarsus in the Early Bronze Age. In *Anatolia and the Ancient Near East: Studies in Honor of Tahsin Özgüç*, eds. K. Emre, B. Hrouda, M. Mellink, and N. Özgüç. Ankara: Tahsin Özgüç'e Armagan, pp. 319-331.

Mellink, M. 1993. The Anatolian South Coast in the Early Bronze Age: The Cilician Perspective. In *Between the Rivers and Over the Mountains: Alba Palmieri Dedicata*, eds. M. Frangipane, H. Hauptmann, M. Liverani, P. Matthiae, and M. Mellink. Rome: University of Rome, pp. 495-508.

Özgüç, T. 1986. New Observations on the Relationship of Kültepe with Southeast Anatolia and North Syria during the Third Millennium B.C. In *Ancient Anatolia: Aspects of Change and Cultural Development: Essays in Honor of Machteld J. Mellink*, eds. J.V. Canby, E. Porada, B.S. Ridgway, and T. Stech. Madison: University of Wisconsin Press, pp. 31-47.

Özten, A. 1989. A Group of Early Bronze Age Pottery from the Konya and Niğde Region. In *Anatolia and the Ancient Near East: Studies in Honor of Tahsin Özgüç*, eds. K. Emre, B. Hrouda, M. Mellink, and N. Özgüç. Ankara: Tahsin Özgüç'e Armagan, pp. 407-418.

Schneider, G. 1988. Stoneware from 3rd Millennium BC: Investigation of a Metal-Imitating Pottery from Northern Mesopotamia. In *Proceedings of the 26th*

International Archaeometry Symposium, Toronto, eds. R.M. Farquhar, R.G.V. Hancock, and L.A. Pavlish. Toronto: University of Toronto, pp. 17-22.

Schneider, G. 1989. A Technological Study of North-Mesopotamian Stone Ware. *World Archaeology* 21 (1): 30-50.

Seton-Williams, M.V. 1954. Cilician Survey. *Anatolian Studies* 4: 121-174.

Shepard, A.O. 1985. *Ceramics for the Archaeologist.* Washington D.C.: Carnegie Institution of Washington.

Tite, M.S., and Maniatis, Y. 1975. Examination of Ancient Pottery Using the Scanning Electron Microscope. *Nature* 257: 122-123.

Tite, M.S., Freestone, I.C. Meeks, N.D., and Bimson, M. 1982. The Use of Scanning Electron Microscopy in the Technological Examination of Ancient Ceramics. In *Archaeological Ceramics*, eds. J.S. Olin, and A.D. Franklin. Washington D.C.: Smithsonian Institution Press, pp. 109-120.

Tite, M.S., Maniatis, Y. Meeks, N.D., Bimson, M., Hughes, M.J., and Leppard, S.C. 1982. Technological Studies of Ancient Ceramics from the Near East, Aegean, and Southeast Europe. In *Early Pyrotechnology: The Evolution of the First Fire-Using Industries*, eds. T.A. Wertime, and S.F. Wertime. Washington D.C.: Smithsonian Institution Press, pp. 61-71.

Vandiver, P., Yener, K.A., and May, L. 1992. Third Millennium B.C. Tin Processing Debris from Göltepe (Anatolia). In *Materials Research Society Proceedings: Materials Issues in Art and Archaeology* 3, vol. 267, eds. P.B. Vandiver, J.R. Druzik, G.S. Wheeler, and I.C. Freestone. Pittsburgh: Materials Research Society, pp. 545-569.

Yener, K.A., and Vandiver, P.B. 1993. Tin Processing at Göltepe, an Early Bronze Age Site in Anatolia. *American Journal of Archaeology* 97: 207-238.

CHAPTER 3

SPECTROSCOPIC CHARACTERIZATION OF ETRUSCAN *DEPURATA* AND *IMPASTO* POTTERY FROM THE EXCAVATION AT PIAN DI CIVITA IN TARQUINIA (ITALY): A COMPARISON WITH LOCAL CLAY

S. Bruni[ß], F. Cariati
Dipartimento di Chimica Inorganica, Metallorganica e Analitica, Università degli Studi di Milano

G. Bagnasco Gianni, M. Bonghi Jovino
Dipartimento di Scienze dell'Antichità, Università degli Studi di Milano

G. Artioli
Dipartimento di Scienze della Terra, Università degli Studi di Milano

U. Russo
Dipartimento di Chimica Inorganica, Metallorganica e Analitica, Università degli Studi di Padova

Some traditional spectroscopic techniques such as Fourier-Transform infrared spectroscopy (FTIR) and near-infrared (NIR) diffuse reflectance spectroscopy have not been fully exploited, as is the case of the former, or even not employed at all, as for the latter, in the investigation of ancient pottery. The present study aims to evidence how these techniques help in characterizing ceramics, especially from the point of view of the firing technology, but also as regards the raw materials and alteration in post-deposition contexts. The mineralogical phases developed in the ceramic material can be individualized through FTIR and subsequently detailed by means of X-ray powder diffraction. It is worth remembering that, for *depurata* ware, petrographic analysis, widely used in the study of coarse ware, results to be scarcely informative. Also, the mineralogical phases, formed when a calcareous illitic-kaolinitic clay, such as that of Tarquinia, is fired, can be related to the maximum temperature reached in the kiln. Diffuse reflectance in the NIR seemed particularly adequate to estimate the firing temperature, allowing to quantify water absorbed by the ceramic material. We tested our method on clay samples from Tarquinia area, fired in our laboratory at temperatures ranging from 500 to 1100 °C. Afterwards we applied it to about one hundred pottery fragments from the excavation at the so-called Pian di Civita. In this way, we verified that: 1) most of the sherds of *depurata* pottery were noticeable homogeneous in their mineralogical features, with differences arising mainly from the firing temperature and, for the more ancient sherds (VIII-VII century BC), from a slightly varying choice of the composition of the raw material; 2) impasto pottery appeared to have been produced from a non-calcareous, sandy material, very rich in feldspars. Finally, a complementary study was conducted by means of Mössbauer spectroscopy to distinguish the oxidation states and coordination environments of iron. In this context, a rather interesting correlation was found with firing temperature.

Traditionally the scientific investigation of archaeological pottery has followed two main lines: on one side, the chemical elemental composition is studied, thus gathering information related to the provenance of pottery itself; on the other side, mineralogical composition is examined, to ascertain not only the production site but also the manufacturing technology of the ceramic artefact. Indeed, the mineralogical phases thus detected are determined by the choice of raw material, by the firing conditions such as temperature and atmosphere and, finally, by the history of the object from its fabrication to its finding in the archaeologist's hands. The method of choice in the study of mineralogical composition is very often petrographic analysis, that, however, while really informative in the case of coarse ware, is of less use when fine ware is under discussion (Mannoni 1994). In the latter case, X-ray powder diffraction (XRD) and also Fourier-transform infrared spectroscopy (FTIR) can prove more useful. It is worth noting that FTIR has not been much exploited in the analysis of archaeological pottery and the present study is also addressed to emphasize its great potential in this sort of investigation. Diffuse-reflectance near-infrared (NIR) spectroscopy and Mössbauer spectroscopy can also be of help in a more precise determination of the firing temperature of an ancient ceramic material, the former informing on the amount of absorbed water and the latter on the chemical environment and oxidation state of iron.

ARCHAEOLOGICAL BACKGROUND

The above-mentioned techniques were used to examine more than 100 samples of Etruscan fine ware from the excavations conducted since 1982 in the so-called area of Pian di Civita in Tarquinia (Viterbo, Italy) (Figure 1), by Prof. Maria Bonghi Jovino. The excavation campaigns brought to light a very important monumental and sacred site of the Etruscan orientalizing period (from the end of the VIII to the first quarter of the VI century BC), to which the name of "complesso sacro-istituzionale" was attributed (Bonghi

[ß] On leave from: *Dipartimento di Scienze Chimiche, Fisiche e Matematiche, Università degli Studi dell'Insubria*

Jovino and Chiaramonte Treré 1997). Given the importance of the site in the Etruscans' religious and every-day life, the archaeological investigation has been supported by a more extensive study of all elements pertaining to the different fields of human activity: in this context, besides paleo-botanical, -zoological and -anthropological analyses, also the examination of the ceramic materials can be used to advantage.

Figure 1. Map showing the location of Tarquinia in the Italian peninsula.

In more details, the specific class of pottery on which the present investigation was mainly focused has been defined as *depurata* ware, its material being characterized by the absence of macroscopically observable inclusions. This class of ceramics was diffused in the whole Mediterranean basin from IX-VIII century BC until the Roman period. The materials from Tarquinia, inside the broader definition of *depurata* pottery, were further grouped in stylistic sub-categories, based on the observation of their superficial decoration (Bagnasco Gianni 1999, in press a, in press b; Sansica 1999), namely:

1) without decoration (plain white ware, *acroma*) (Fig. 2);
2) with a stripe decoration, or banded (*a bande*) (Fig. 2);
3) with geometrical decorative motifs (Etrusco-geometric, *etrusco-geometrica*);
4) with Corinthian-style decorative motifs (Etrusco-Corinthian, *etrusco-corinzia*) (Fig. 3);
5) archaic black-glazed (*a vernice nera arcaica*).

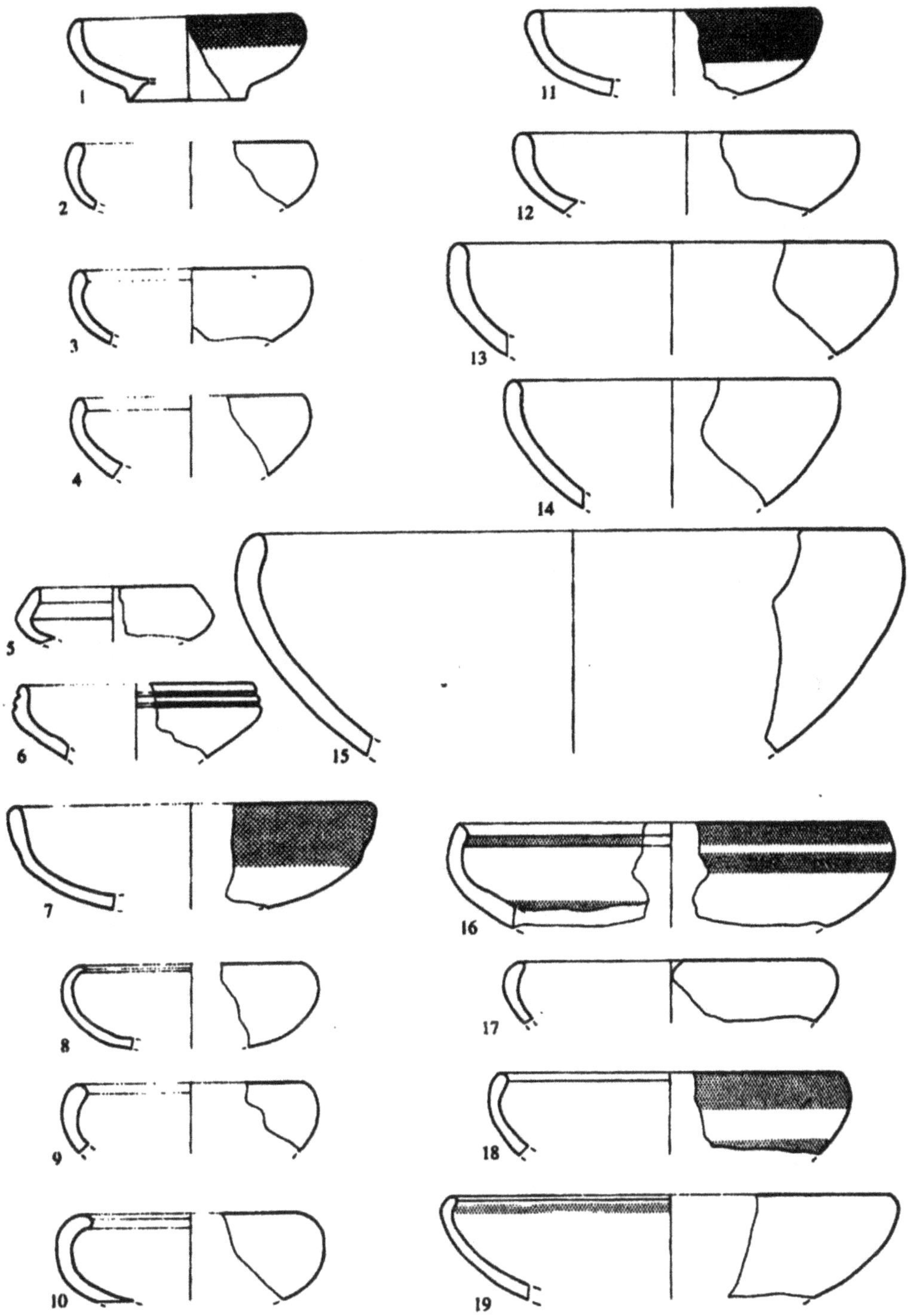

Figure 2. Examples of plain white ware and banded pottery from Pian di Civita (half-scale).

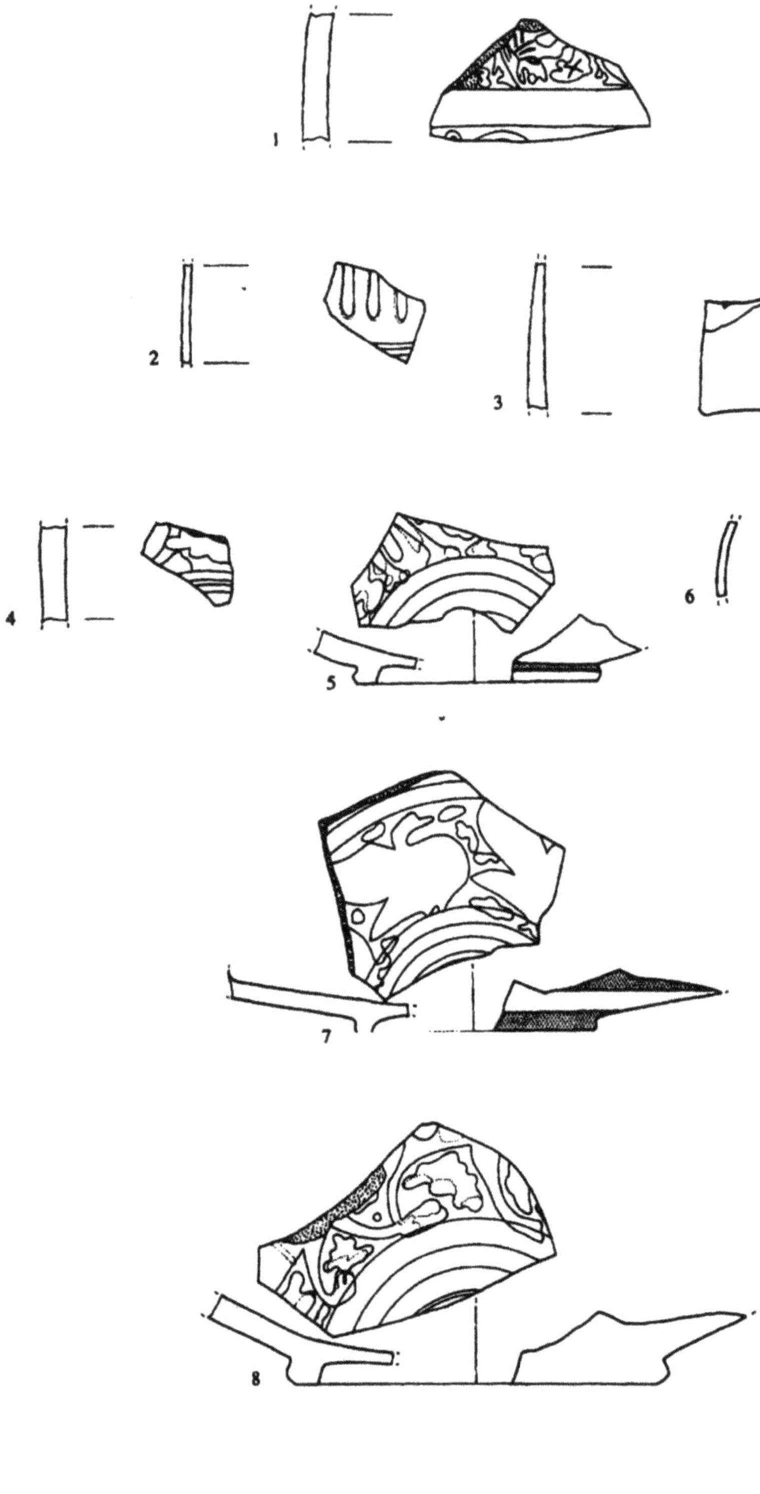

Figure 3. Examples of Etrusco-Corinthian pottery from Pian di Civita (half-scale).

Once established on the basis of elemental composition that most of the examined samples form a homogenous group, thus pointing to a common local origin (Cariati et al. in preparation), the remaining problem to be discussed was indeed the manufacturing process, in terms of selection and treatment of the raw material and of firing conditions.

For comparison samples of clay from the Pian di Civita were also collected, treated as described below and examined by means of the same techniques used for the ceramic sherds. It is worth remarking that, while ancient pottery workshops were not found at Pian di Civita, clay from that area is used by modern craftsmen in Tarquinia to imitate Etruscan vases.

RESULTS AND DISCUSSION

XRD, FTIR and diffuse-reflectance NIR spectroscopies
In the ternary phase diagram CaO(+MgO)-Al_2O_3-SiO_2 (Fig. 4) most of the ancient fine ceramic materials are located in the quartz-anorthite-diopside (or wollastonite) compatibility triangle (Heimann and Maggetti 1981; Heimann 1989) and the samples under discussion are no exception to the rule.

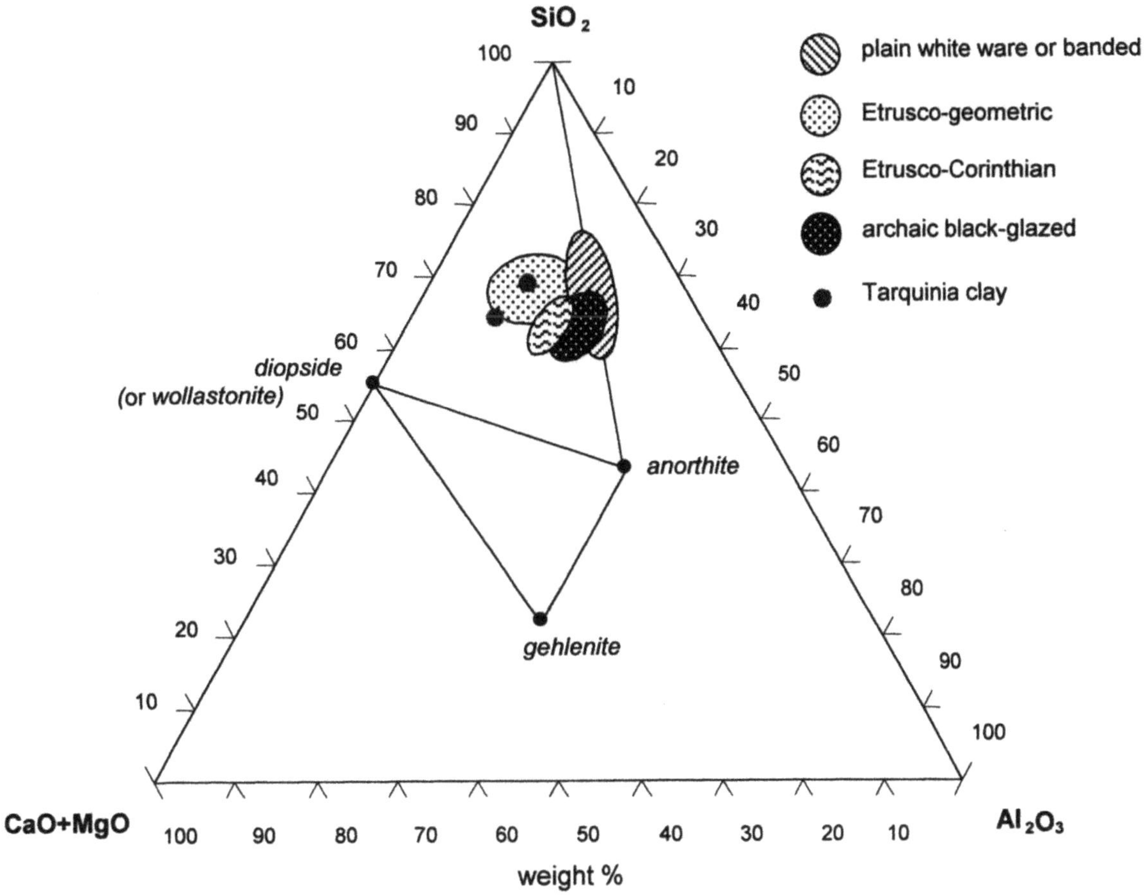

Figure 4. Composition of Etruscan *depurata* pottery from the excavation of Tarquinia in the ternary system CaO-Al_2O_3-SiO_2.

However, the mineralogical composition of a ceramic sample is that expected on the basis of the diagram in Figure 4 only if the starting material was homogeneous and if the firing temperature was high enough to ensure that the conditions of thermodynamic equilibrium were reached (Heimann 1989). If the raw material was coarse or if the firing temperature did not allow a complete melting of the mixture, -and this is what usually happened- different mineralogical phases can be encountered in the ceramic. Various studies (Heimann and Maggetti 1981; Heimann 1989; Letsch and Noll 1983) report on the correlation between the mineralogical phases developed upon firing a calcareous illitic-kaolinitic clay, such as Tarquinia clay, and the firing temperature. It is well established, then, that: (i) around 800 °C the decomposition of calcium carbonate ($CaCO_3$) begins, forming calcium oxide (CaO); (ii) at temperatures above 850-900 °C CaO can react with Al_2O_3, MgO and SiO_2, delivered by illite, kaolinite and dolomite, forming diopside ($CaMgSi_2O_6$) or wollastonite ($CaSiO_3$) and gehlenite ($Ca_2Al(AlSi)O_7$); (iii) the latter is clearly a metastable phase in the examined compositional domain (Figure 4) and therefore it transforms into anorthite ($CaAl_2Si_2O_8$) above 1050 °C. Heimann and Maggetti (1981) also discussed the formation of mineralogical phases due to the alteration, taking place during the burial of the artefact, of the phases previously formed upon firing of the artefact itself. For example, secondary calcite can derive from the decomposition of gehlenite in those sherds buried in humid environments.

Figure 5 (*a*) and (*b*) shows the XRD patterns and the FTIR spectra of test samples made from Tarquinia clay, having the composition indicated in the phase diagram above (Fig. 4), worked with water and subsequently fired in a laboratory electrical furnace at a maximum temperature ranging from 500 °C to 1100 °C for 120 minutes. As expected, the typical signals of calcite disappear at temperatures higher than 800°C, while illite and kaolinite decompose, forming an amorphous silicate matrix and calcium silicates and aluminosilicates, such as wollastonite, diopside, gehlenite and anorthite. Instead of an equilibrium assemblage, large amounts of gehlenite are formed and maintained even for a maximum firing temperature as high as 1100 °C. It should be observed that, even if the usual procedure in this experiment was to keep the samples at the highest temperature of each firing cycle for two hours, also prolonging the isotherm at 1100 °C to six hours nothing was substantially changed in the resulting mineralogical composition of the test sample. Therefore, the presence of a considerable quantity of gehlenite can be associated neither with the mean elemental composition of the raw material, as shown in Figure 4, nor with the firing conditions, but rather with the grain size of the raw material itself. Indeed, in the experiment described, clay was used as such, without any depuration treatment, and it has been suggested (Heimann and Maggetti 1981; Heimann 1989) that a coarser granulometry should favour the formation of gehlenite as an effect of the different local composition at the grain boundaries.

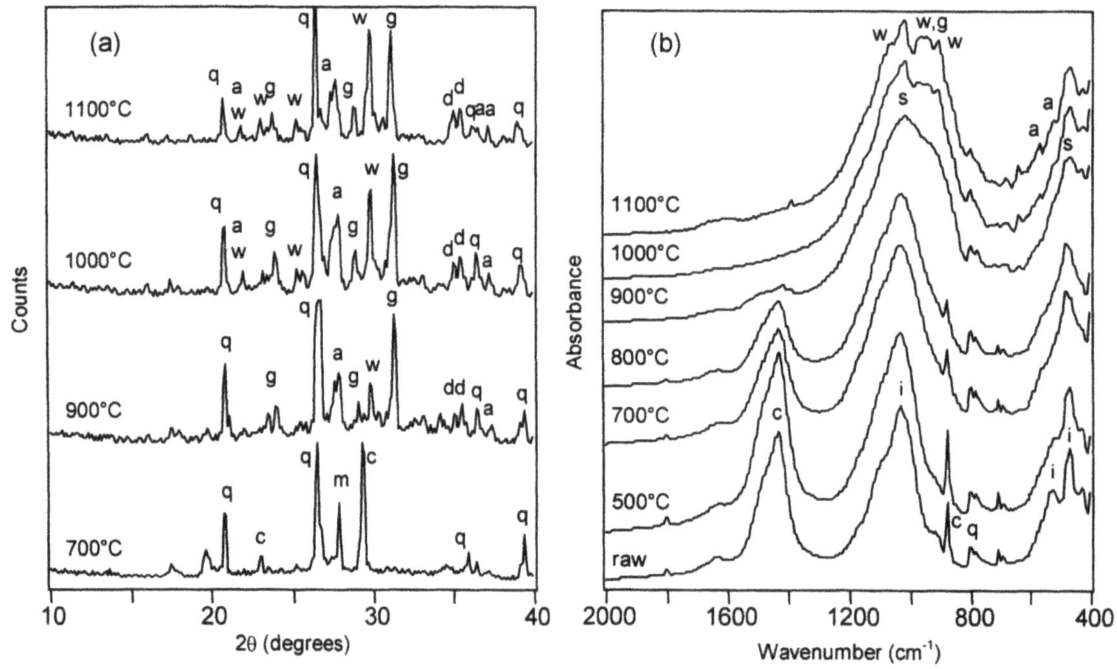

Figure 5. (a) X-ray diffractograms and (b) FTIR spectra of test samples of Tarquinia clay heated at different temperatures. q=quartz, c=calcite, w=wollastonite, g=gehlenite, d=diopside, a=anorthite, i=illite, s=disordered silicate network.

From the methodological point of view, Figure 5 shows the effectiveness of the XRD and FTIR techniques in following the changes of the mineralogical composition of clay upon firing. Therefore, the same techniques were applied to the study of samples of plain white ware and banded pottery, dating from VII to VI century BC and later. As shown in Figure 4, the samples are located in a rather restricted compositional range. Indeed the differences in the observed mineralogical phases can be related to the different firing temperatures, following the above outlined scheme, as can be seen in Figure 6 (opposite page).

Therefore, in going from sample 3/1010 to sample 3/1005, a firing process of increasing temperature can be assumed, so that if sample 3/1010 was not heated above 800°C, as the presence of significant amounts of not decomposed calcite and illite demonstrates, the development of increasing quantities of diopside and anorthite suggests a firing temperature of at least 850°C for sample 170/78, while the presence of much anorthite and the complete disappearance of gehlenite indicates that a temperature around 1000 °C was probably reached by sample 3/1005 when it was fired. For sample 3/1002, the characteristic signals of calcite are observed as well as those of diopside, gehlenite and anorthite, indicating the presence of secondary calcite, probably deriving from the partial decomposition of gehlenite due to humid burial as described above.

Some differences in type and relative quantities with the phases detected in the test samples of Tarquinia clay can be noted: (i) first of all, none of the samples of plain white ware or banded pottery shows so large amounts of gehlenite as those found in the test samples of fired clay; this simply indicates a skilful processing of the raw material, aimed to render it sufficiently fine-grained avoiding large quartz grains; (ii) diopside is prevalently formed with respect to wollastonite, suggesting a larger percentage of magnesium in the clay used by the ancient potters; and (iii) the ratio anorthite/diopside at the highest firing temperatures clearly favours anorthite in the Etruscan samples, but this is in agreement with the different location in the phase diagram of Figure 4 of the modern clays and ancient ceramics. The last two factors confirm once more an observation that is recurrent in the literature, i.e. that of the difficulty to find nowadays samples of clay perfectly comparable with those used in the antiquity by potters in a given production site

(Farnsworth 1970; Dupont 1977, 1983). The deposits exploited in ancient times might not be easily accessible at present or, even more probably, craftsmen used mixtures of different clays to ensure the best properties of the ceramic material (Heimann 1989).

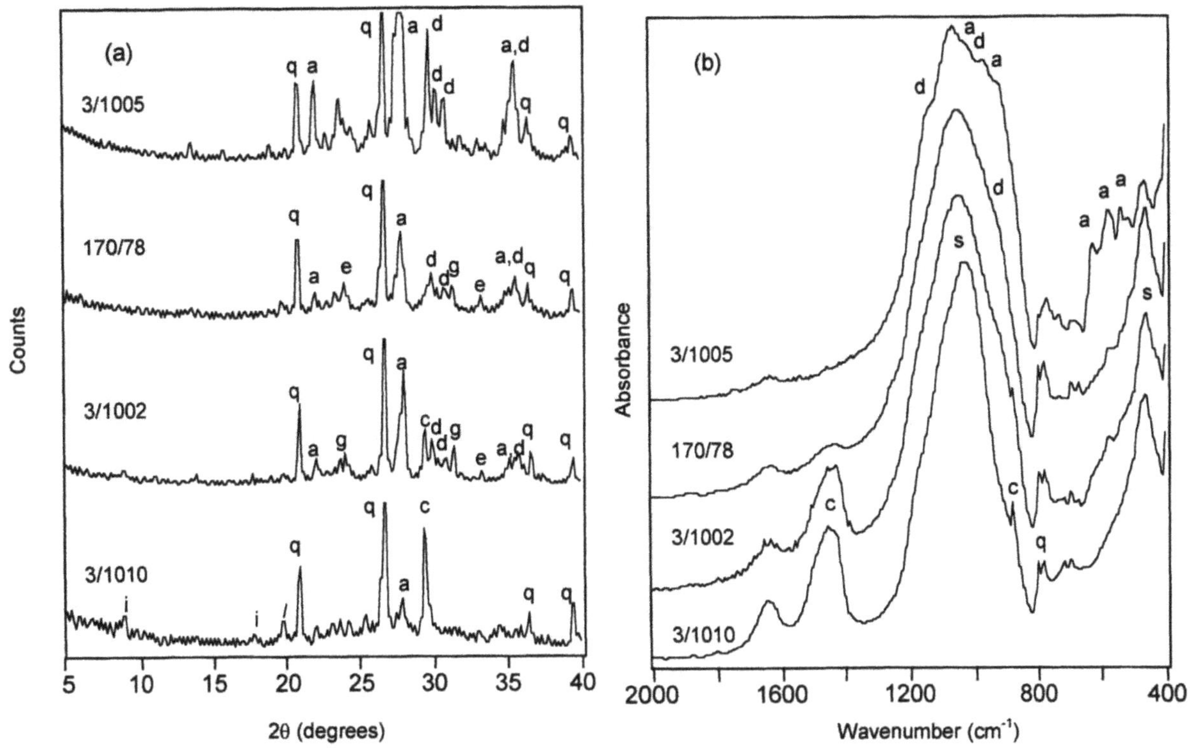

Figure 6 (a) X-ray diffractograms and (b) FTIR spectra of samples of Etruscan *depurata* plain white ware and banded pottery (VII-VI century onwards). For symbols, see Figure 5.

One way independent from the mineralogical composition to estimate the relative, if not absolute, firing temperature of pottery sherds, is by evaluating the amount of absorbed water from the intensity of the corresponding signal in the near-infrared (NIR) region of the electromagnetic spectrum. In Figure 7, the diffuse-reflectance NIR spectra of the fired test samples of Tarquina clay are reported. In these spectra, the intensity of the band assigned to absorbed molecular water (Cariati et al. 1981,1983) undergoes an evident decrease upon increasing the firing temperatures of the test samples. Indeed, if the diffuse-reflectance NIR spectra of the same ancient sherds examined above are considered (Figure 8), the variation of the amount of absorbed water results to be consistent with the previously suggested hypotheses concerning the maximum temperature they reached when fired.

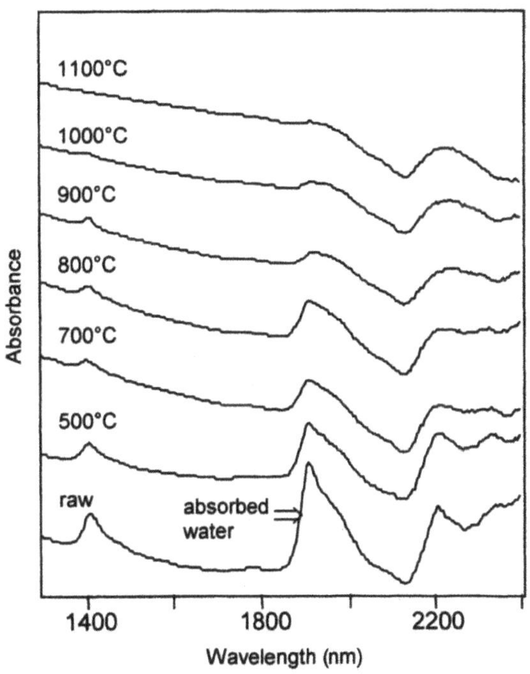

Figure 7. Diffuse-reflectance NIR spectra of test samples of Tarquinia clay fired at different temperatures.

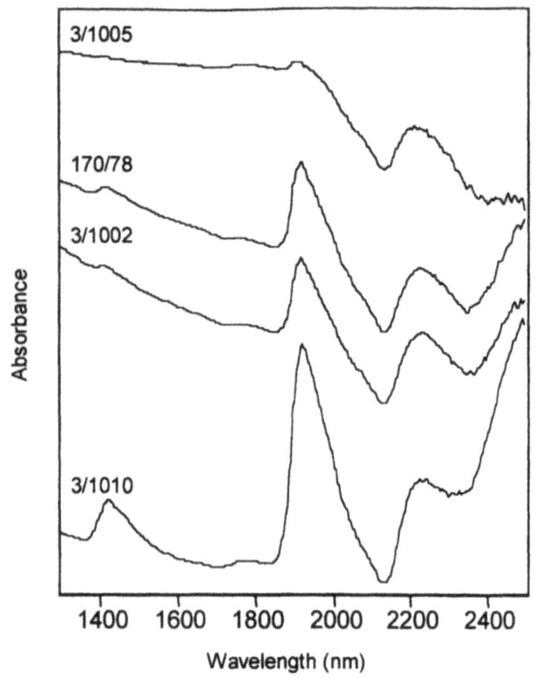

Figure 8. Diffuse-reflectance NIR spectra of samples of Etruscan *depurata* plain white ware and banded pottery.

It is worth noting that: (i) samples 170/78 and 3/1002 were in fact fired at similar temperatures, and the calcite detected in the latter is therefore confirmed to be of secondary origin; (ii) the amount of water absorbed in a sherd for which a firing temperature as high as 800-900 °C is estimated on the basis of the mineralogical composition is greater than that found in a test sample of clay heated in a laboratory electrical furnace at the same temperature. This last fact can be easily explained by considering that ancient materials could reach an equilibrium content of water, i.e. they had time to absorb as much water as the sites available in their silicate network allowed them to in the burial environment, while this is not the case for the recently fired test samples.

On the basis of the above-discussed criteria, sherds of Etrusco-geometric (VIII-VI century BC), Etrusco-Corinthian (VIII-VI century BC) and black-glazed (VI century BC onwards) pottery were examined. Examples of X-ray diffractograms and FTIR spectra are shown in Figure 9 for a representative sample for each class. The experimental data demonstrate that the most ancient samples, i.e. the Etrusco-geometric ones (sample 88/74 in Figure 9), are characterized by a considerable content of gehlenite, generally higher than that found for plain white ware and banded pottery, and variable relative amounts of diopside and anorthite. The quantity of absorbed water (Figure 10) is usually comparable with that present in those samples of plain white ware and banded pottery that were presumably fired between 800 and approximately 900°C . The large amount of gehlenite, as stated above and as evidenced by the firing experiments of test samples of clay, is related to the coarse granulometry of the raw material and also causes the presence of reformed calcite; in parallel, the variability in the ratio anorthite/diopside seems to depend on the grain size of primary calcite (Heimann and Maggetti 1981; Heimann 1989).

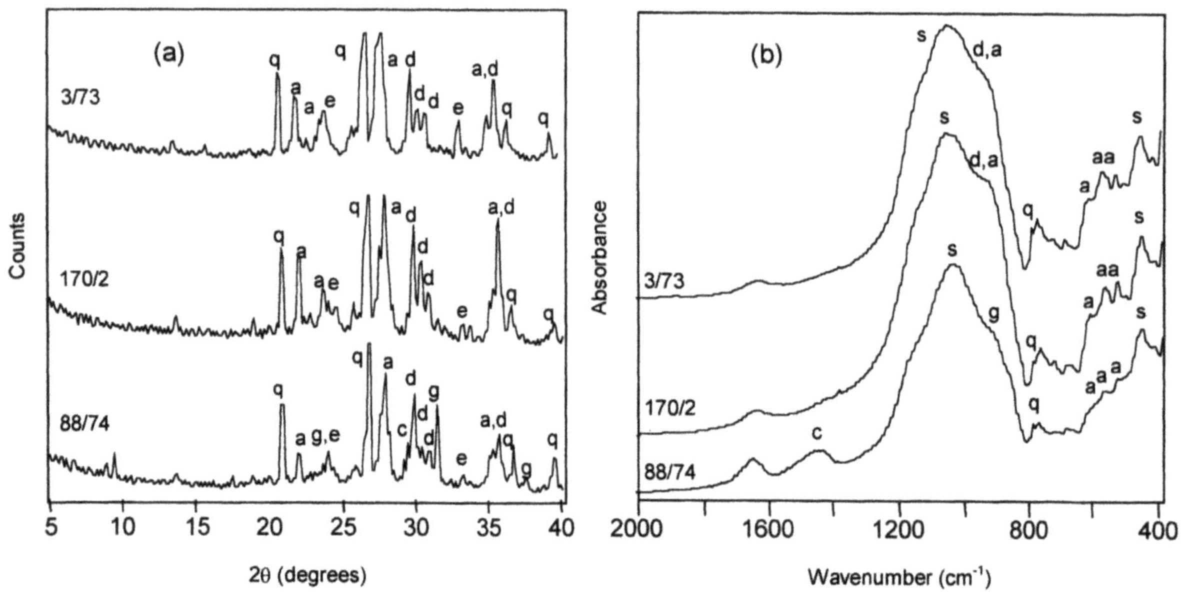

Figure 9 (a) X-ray diffractograms and (b) FTIR spectra of sample 88/74 (Etrusco-geometric pottery, VIII-VII century BC), sample 170/2 (Etrusco-Corinthian pottery, VI century BC) and sample 3/73 (archaic black-glazed pottery, VI century BC onwards) from the excavations of Tarquinia. For symbols, see Figure 5.

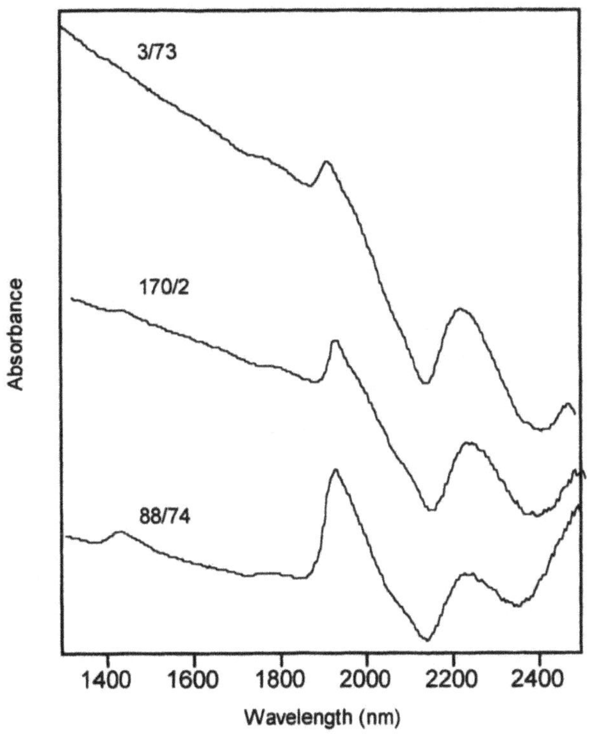

Figure 10. Diffuse-reflectance NIR spectra of sample 88/74 (Etrusco-geometric pottery, VIII-VII century BC), sample 170/2 (Etrusco-Corinthian pottery, VI century BC) and sample 3/73 (archaic black-glazed pottery, VI century BC onwards) from the excavations of Tarquinia.

These observations are in agreement with the earlier chronology of Etrusco-geometric pottery with respect to the other classes of *depurata* pottery, suggesting a poorly developed technology, both in the choice and pre-treatment of the raw material, less precise also from the point of view of elemental composition (see the quite wide compositional range of Etrusco-geometric samples in Figure 4), and in the difficulty in obtaining high firing temperatures. On the contrary, both Etrusco-Corinthian and black-glazed sherds are quite similar, as far as the mineralogical composition and the amount of absorbed water are concerned, to the samples of plain white ware and banded pottery that resulted to have been fired at temperatures from 950 to over 1000 °C: the X-ray diffractograms and FTIR spectra of sample 170/2, for Etrusco-Corinthian pottery, and of sample 3/73, for archaic black-glazed pottery, show clearly (Figure 9) that anorthite and diopside were formed in significant proportions upon firing and that the raw material was fine enough and the firing temperature sufficiently high to limit the formation of gehlenite. Coherently, the water content (examples in Figure 10) is from medium to low for the Etrusco-Corinthian materials and reproducibly low for all the black-glazed sherds. In the manufacturing of these last materials, the need to obtain a well-vitrified gloss obliged the potters to attain high firing temperatures (Jones 1986) and the reported results indicate that at least from the VI century on, the craftsmen in Tarquinia possessed a rather specialized technology, allowing them to plan and realize a given firing cycle, minimizing also the effects on the maximum temperature experienced by a vase as a consequence of its different location in the furnace.

For comparison, some samples of *impasto* pottery, again from the excavations of the Civita and dating from X to VII century BC, were also investigated by means of the XRD, FTIR and diffuse-reflectance NIR techniques. The mineralogical composition of these sherds is, as expected, completely different from that of *depurata* pottery, being characterized by an abundant content of sodium and potassium feldspars and by the absence of calcite, in spite of a rather low firing temperature demonstrated by the appreciable amount of absorbed water. The hypothesis of a sandy rather than clayey raw material seems reasonable on the basis of these data.

Mössbauer spectroscopy

A further confirmation of the firing temperature range determined for the *depurata* pottery from the Civita of Tarquinia was obtained by means of Mössbauer spectroscopy. For almost all samples, except some of the black-glazed samples having a grey colour, iron is mostly found in its highest oxidation state, i.e. as iron(III). Small percentages of iron(II) are also possibly found, around 10-15 percent. Iron(III) gives rise in some samples to a magnetic component, represented by a sextet in the Mössbauer spectrum, and, in all the sherds examined by means of this technique, to one or, more frequently, two paramagnetic components, appearing as doublets.

Mössbauer spectra were collected also for the fired test samples of Tarquinia clay. For such samples, when fired at temperatures below 900 °C, the two components are due to iron(III) in environments respectively described as $FeO_{6-x}(OH)_x$ and FeO_6, while, for higher firing temperatures, they are related to iron(III) in the amorphous silicate network and in the crystalline lattice of those phases developed upon firing, such as diopside. If the quadrupole splittings ΔE_Q, i.e. the distances between the two signals of each doublet, are plotted against the firing temperature, a correlation similar to that discovered by other researchers for clays of different provenance (Maniatis et al. 1981; Deriu 1989) is obtained, as shown in Figure 11: a maximum is reached around 700 °C and then the values of the quadrupole splittings decrease until 1000 °C, starting to rise again above this temperature. When the quadrupole splittings of the ferric doublets are plotted in the same graph for a number of samples of *depurata* pottery from the excavations of Tarquinia, referring to the firing temperature estimated on the basis of the mineralogical phases and the amount of absorbed water, an excellent correspondence is found, as it is apparent in Figure 11.

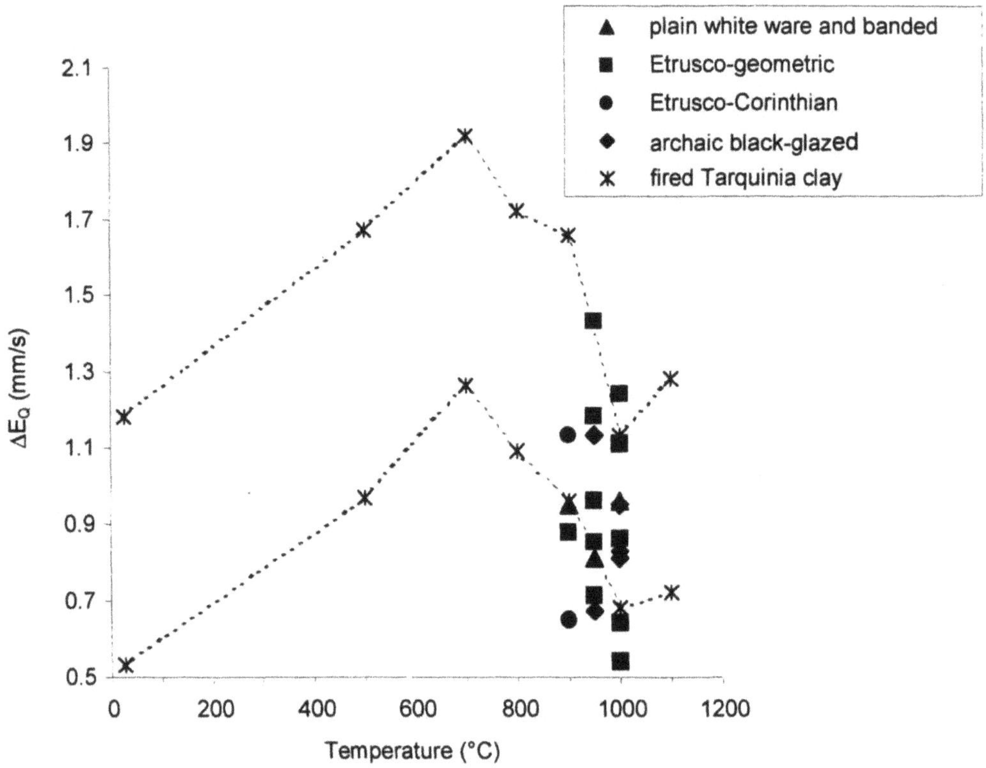

Figure 11. Variation of quadrupole splitting with firing temperature for the paramagnetic doublets of the room-temperature Mössbauer spectra of test samples of Tarquinia clay (- ✱ -). Symbols indicate the same parameters for the samples of *depurata* pottery from the excavations of Tarquinia, referred to the firing temperatures hypothesized on the basis of the mineralogical composition and water content.

CONCLUSION

In conclusion, the study here described represents an interesting application of conventional spectroscopic techniques to the investigation of the manufacturing technology of Etruscan fine ware from the site of the Civita in Tarquinia. The obtained results indicate clearly an evolution from VIII to VI century in the potters' skill, represented mainly by a more precise control of the composition of the raw material and by the ability to reach and maintain quite high temperatures upon firing. The histogram reported below (Figure 12) summarizes this evolution from the point of view of the firing temperature determined for the different ceramic classes.

The results obtained from the spectroscopic analysis match quite well the historical reconstruction made on archaeological bases. As stated above, the production of *depurata* ware can be tracked in Tarquinia from the first half of the VIII to the whole II century BC. More in detail, during the first half of the VIII century the production of the Etrusco-geometric pottery began, giving rise to vessels homogenous from the stylistic point of view, even if characterized by a quite variable ceramic paste. The analyses here summarized indicate the cause of this variability in a raw material naturally rather fine-grained but not depurated further on.Starting from the end of the VII century BC, the situation began to change with the onset of the production of Etrusco-Corinthian pottery and, contemporarily, of East-Greece fashion banded ware. Under the influence of the East Greek world, an increase of the production rate also took place besides the stylistic evolution. A stabilization of the manufacturing technique, as evidenced from the above discussed spectroscopic results, was an obvious part of this picture, and was completely fulfilled when, from the middle of the VI century, black-glazed pottery began to be produced, requiring a well-established skill to obtain not only a satisfying ceramic paste but also a properly coated surface. There are no elements allowing to suppose the presence in Tarquinia of different workshops specialized in the production of a given class of ware. It seems more reasonable to hypothesize that, from the VI century on, a certain number of workshops or potters were able to manufacture vessels of different style, adjusting their technique to the requirements of the finished product.

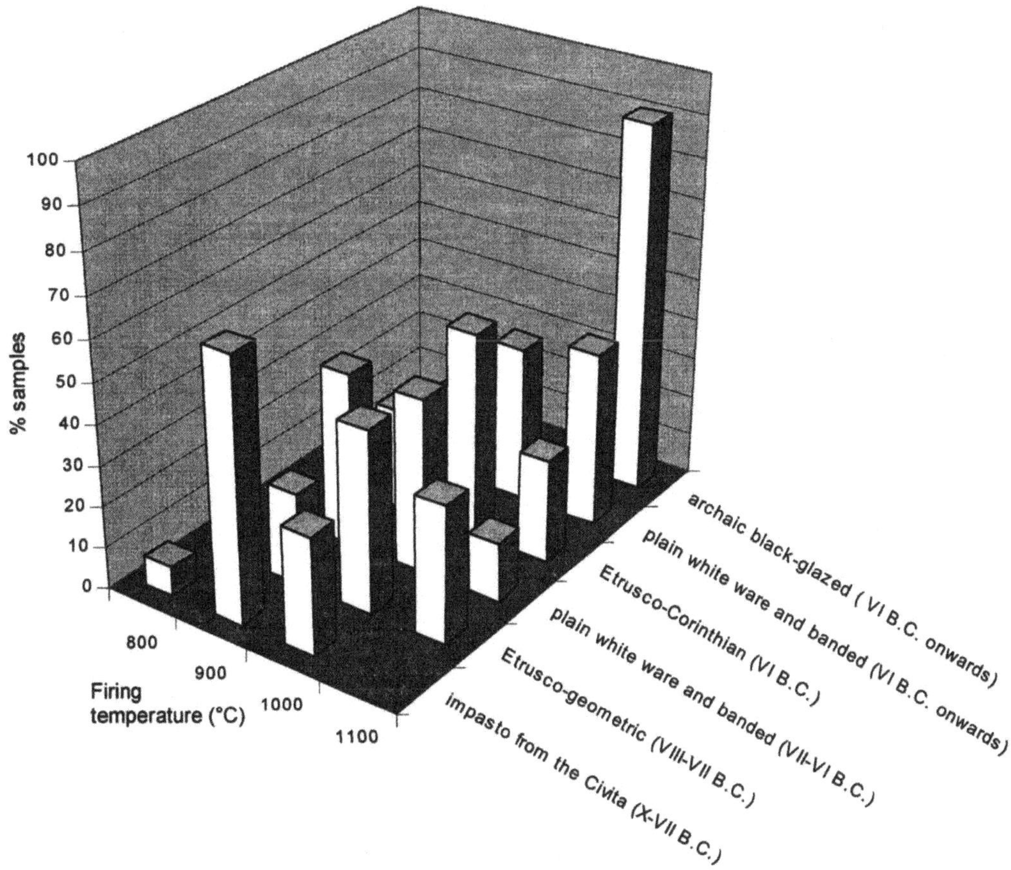

Figure 12. Percentages of samples of different classes of *depurata* pottery and of *impasto* pottery as a function of the estimated firing temperature.

Acknowledgments

The Consiglio Nazionale delle Ricerche CNR is acknowledged for financial support. The authors wish to thank the Soprintendenza Archeologica per l'Etruria Meridionale, the Municipality of Tarquinia and the Università Agraria di Tarquinia for co-operation in the archaeological researches.

References

Bagnasco Gianni, G. 1999. La ceramica depurata acroma e a bande. In *Tarchna II. Tarquinia. Scavi sistematici nell'abitato. Campagne 1982-88. I materiali*, ed. C. Chiaramonte Treré. Roma: L'Erma di Bretschneider, pp. 99-176.

Bagnasco Gianni, G. in press a. La ceramica di importazione e etrusco-geometrica. In *Tarchna III. Tarquinia. Scavi sistematici nell'abitato. Campagne 1982-88. I materiali*, ed. M. Bonghi Jovino. Roma: L'Erma di Bretschneider.

Bagnasco Gianni, G. in press b. La ceramica etrusca a vernice nera arcaica. In *Tarchna III. Tarquinia. Scavi sistematici nell'abitato. Campagne 1982-88. I materiali*, ed. M. Bonghi Jovino. Roma: L'Erma di Bretschneider.

Bonghi Jovino, M. and Chiaramonte Treré, C. 1997. *Tarquinia. Testimonianze archeologiche e ricostruzione storica. Scavi sistematici nell'abitato. Campagne 1982-1998*. Roma: L'Erma di Bretschneider

Cariati, F., Erre, L., Micera, G., Piu, P., and Gessa, C. 1981. Water molecules and hydroxyl groups in montmorillonites as studied by near infrared spectroscopy. *Clays and Clay Minerals* 29, 157-159.

Cariati, F., Erre, L., Micera, G., Piu, P., and Gessa, C. 1983. Effects of Layer charge on the near-infrared spectra of water molecules in smectites and vermiculites. *Clays and Clay Minerals* 31, 447-449.

Cariati, F., Fermo, P., Bagnasco Gianni, G., Todeschini, R., and Consonni, V., in preparation. (2001). A provenance study of *depurata* and *impasto* ceramic from Tarquinia (Italy).

Cuomo di Caprio, N. 1985. *La Ceramica in Archeologia*. Roma: L'Erma di Bretschneider.

Deriu, A. 1989. Caratterizzazione di ceramiche greche e campane dell'VIII sec. a. C. mediante spettroscopia Mössbauer. In *Atti del Secondo Congresso Nazionale Etrusco*. Roma: Giorgio Bretschneider, pp. 79-91.

Dupont, P. 1977. Recherches de laboratoire sur les céramiques grecques orientales archaiques. *Revue d'Archéométrie* 1, 105-114.

Dupont, P. 1983. Classification et détermination de provenance des céramiques grecques orientales archaiques d'Istros. *Dacia* 27, 19-43.

Farnsworth, M. 1970. Corinthian pottery: technical studies. *American Journal of Archaeology* 74, 9-20.

Heimann, R. B. 1989. Assessing the technology of ancient pottery: the use of ceramic phase diagrams. *Archaeomaterials* 3, 123-148.

Heimann, R. B., and Maggetti, M. 1981. Experiments on simulated burial of calcareous terra sigillata (mineralogical change). Preliminary results. *Occasional Papers, British Museum* 19, 163-177.

Jones, R. E. 1986. *Greek and Cypriot Pottery*. Athens: The British School at Athens.

Letsch, J., and Noll, W. 1983. Mineralogie und Technik der frühen Keramiken Thessaliens. *Neues Jahrbuch für Mineralogie, Abhandlungen* 147: 109-146.

Maniatis, Y., Simopoulos, A., and Kostikas, A. 1981. Moessbauer study of the effect of calcium content on iron oxide transformations in fired clays. *Journal of the American Ceramic Society* 64, 263-269.

Mannoni, T. 1994. *Archeometria. Geoarcheologia dei manufatti*. Genova: ESCUM.

Sansica, G. 1999. La ceramica depurata etrusco-corinzia. In *Tarchna II. Tarquinia. Scavi sistematici nell'abitato. Campagne 1982-1988. I materiali*, ed. C. Chiaramonte Treré. Roma : L'Erma di Bertschneider, pp. 177-204.

CHAPTER 4

PETROLOGICAL EDS CHEMICAL STUDY IN THIN SECTION OF SOME ETRUSCO-CORINTHIAN CERAMICS: A CONTRIBUTION TO THEIR ARCHAEOLOGICAL KNOWLEDGE.

Gérard Bossière
Laboratoire de Planètologie et Géodynamique, Faculté des Sciences et des Techniques de Nantes, France (UMR CNRS 6112)

Dominique Frère
Université de Bretagne Sud, UFR d'Histoire, Lorient, France, ETUMR CNRS 126-4

This paper presents SEM-EDS analysis of five thin sections of Etrusco-Corinthian ceramics of the Recent Orientalisant, from excavations in Calatia (Italy). Analyses were made on the cement and on mineral grains included in the ceramics. They have allowed us to (1) determine the mineralogical and chemical characteristics of each sample; (2) show that sample ET4, which is known to be made locally, has a specific composition corresponding to a greywacke. The original clay for sample ET4 has a different source than the other samples. We emphasize two points. First, in two ceramic samples, volcanic pyroxenes were found associated with phyllosilicates (white micas), minerals, which have quite different paragenesis. This indicates that pyroxene was added, probably from a sand, to the clay base; consequently, pyroxene must be considered as part of the temper added to the clay. Secondly it is shown that melting sometimes occurred during firing of these ceramics, sometimes in association with carbon inclusions, sometimes not.

INTRODUCTION

Few scientists have studied the chemistry and mineralogy of Etrusco-Corinthian wares. Bruni et al. (this volume) are among them. However, contrary to the study by Silvia Bruni and her colleagues working on very fine pastes using Fourier-Transform infrared spectroscopy and near-infrared diffuse reflectance spectroscopy, our analysis focuses on the mineralogical grains in the matrix using a scanning electron microscope coupled to energy dispersive spectrometer.

The Beazley's method (Spivey 1988) is the only one which has been routinely used in history of art for stylistic identification and classification of thousands of vases and fragments dating from the last third of VIIth century and the first half of the VIth century BC (Szilàgyi 1992; Frère 1995). This important phase of production began in south Etruria and extended not only to various sites of central Italy but also in Campania. In this area, Calatia, Capua and Pontecagnano (Figure 1) are the cities with the greatest quantity of imported Etruscan ceramics during the Recent Orientalisant (Colonna 1961, 1964; Albore Livadie 1979; Meyer 1993).

The discovery of imported Greek or Etruscan ceramics, is always of importance to the archaeologist and the historian for relative dating and information on trade in western Mediterranean. However, confusion caused by locally produced imitations of the Greek or Etruscan ceramics often introduce errors in the interpretations of the material from a necropolis.

Therefore, it is of main importance to distinguish imported ceramics from locally made ones (Druc 2000; Shriner and Murray 2000). A first set of SEM Tracor EDS chemical analyses of one sample (Bossière and Moguedet 1995) allowed us to determine that (1), the cement of the ceramic may be classified between «low-alumina fireclay» and «fireclay» similar to the so-called pozzolanas; (2), the temper is of different mineralogy (amphibole, pyroxenes, phyllosilicates, alkali feldspars, al-silicate), and (3), the pyroxenes, which were probably added to the clay as temper, are similar to those found in alkaline volcanics.

The goal of this paper is to study, from a petrological point of view, the compositional differences in the cement and temper, in order to distinguish the production of perfume vases from southern Etruria and those from central Campania (Frère 1995). The method used was EDS analysis with a SEM following examination of the section in the backscattered electron mode (BSE); analyses were performed on the fine matrix and on the coarser minerals grains set in the matrix. Analyses of the matrix were studied with the aid of ternary diagrams and changes in the mineral chemistry of coarser grains, from sherd to sherd, were compared.

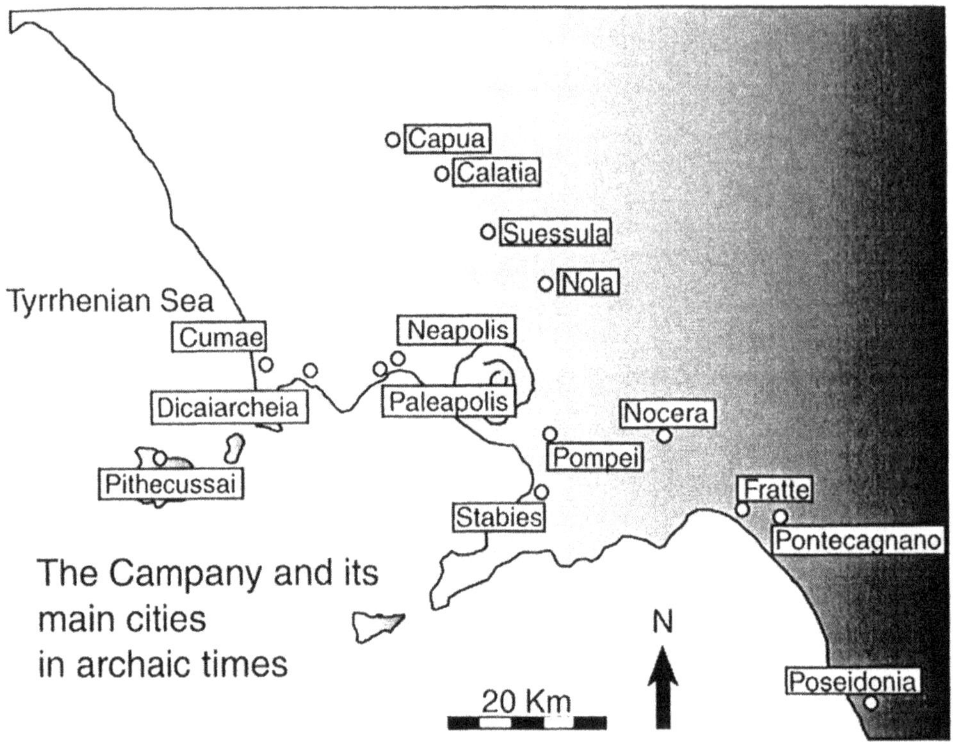

Figure 1. Map of the area of study.

ORIGIN OF THE CERAMICS STUDIED

The analyses were performed on five ceramic sherds, ET1 to ET5. They date from the Recent Orientalisant Period (625-575 BC), and come from excavations in Calatia (Table 1, Figure 1). The sherds come from four perfumed-oil vases, aryballoi and alabastroi, and one drinking cup (Figure 2). ET2 may either be imported from Etruria or locally made. ET4 must be a local production from the first or second quarter of the 6th century, since, there was no more Etruscan importation at this time. These vases are in sub-geometric or in black figured decoration (Payne 1931; Neeft 1987; Amyx 1988; Szilàgy 1992, 1998; Belleli 1997; Boardman 1998). The vases are of small dimensions (between 4 - 25 centimeters height), wheel-turned, and are made to contain and to pour perfumed oil. They have a very fine paste, hard (the color is white pinkish or orange) and a lightly shiny black-brown glaze.

Analytical data from sample Nimes 863-1-21 (Bossière and Moguedet 1995) were added for comparison. This sample comes from Caere, 50 km north of Roma, 10 km inland. It is considered to have been made locally.

Table 1. Type and estimated origin of the samples analyzed based on stylistic studies and aspect of the ceramic and glaze.

Samples	**ET1**	**ET2**	**ET3**	**ET4**	**ET5**
Tombs	Calatia 140	Calatia 324 ANAS n°11	Calatia 277 n°52	Calatia 324 n°18	Calatia ANAS 304
Types	Pyriform aryballoi	Drinking cup	Globular aryballoi (bird's cycle)	Globular aryballoi	Alabastroi
Dating	circa 640-620 BC	circa 600-570 BC	circa 570-550 BC	circa 580-560 BC	circa 620-590 BC
Geographic origin	southern Etruria	southern Etruria OR inland Campania	southern Etruria	Inland Campania	southern Etruria

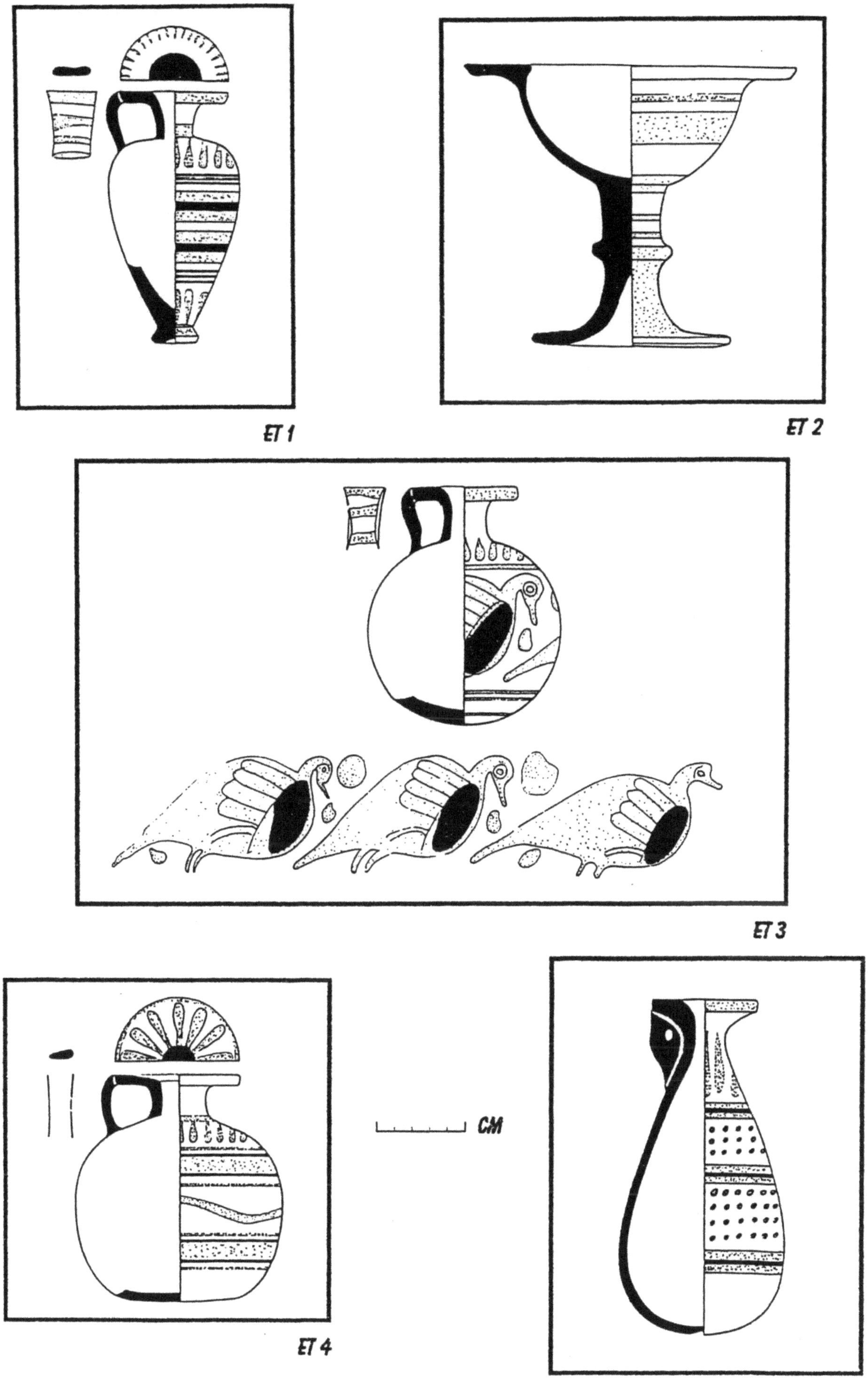

Figure 2. Illustration of the samples studied.

EXPERIMENTAL METHOD

SEM examination of the ceramic samples allows one to distinguish three different parts or phases: 1- voids, 2- a very fine-grained cement around the voids, originally made of clay, referred to as the matrix in this paper and, 3- minerals (muscovite, biotite, quartz, feldspars, pyroxenes, amphiboles) which may have been originally in the clay or added as constituents of a second material (or temper).

Bulk chemical analysis typically does not distinguish the matrix from the minerals. Thus, if some minerals were added as temper, the analysis would not be representative of the original clay. In some cases, fluxing agents or other materials (P_2O_5 for example, see Bossière and Moguedet 1995, or Velde and Druc 1999 for a review) may also be added. This study is based on the fact that it is the clay (or an association clay-fluxing agent) that yields the ceramic, once fired. So spot analysis was made on the cement (or matrix) constituting anastomosed trabecula (Figure 3).

Semi quantitative analysis (SSQ) was performed on polished, carbon coated thin sections by a EDS Tracor system (TN 5500), on a SEM microscope JEOL JSM 35C. SSQ analysis uses its own standards corresponding to pure elements. The microscope operates at an excitation voltage of 15KV, with an image current of 0,1 nA, and time counts between 40 to 60 seconds. The electronic beam diameter is 1 nm.

Minerals and melted areas were analyzed quantitatively using a JEOL 5800 LV microscope under the same experimental conditions as described above. The EDS system was a PGT (IMIX-PTS model). The time count was 60 seconds and the analyses were calculated using external natural standards.

Previous to the analysis, the use of the backscattered electron mode (BSE) allows one to ensure the analyzed zone is homogeneous. One must keep in mind that the volume, which is analyzed, has a pear shape whose bulbous part is situated under the analyzed surface, and which, therefore, is not visible. A mineral may be hidden under the surface, affecting the accuracy (for example, a quartz grain situated underneath the surface will give a silica weight percent too high for a given mineral). It is expected that multiple analyses solve this problem, by working on a population and avoiding non-representative analysis

Figure 3 shows backscattered electron (BSE) photographs of the samples studied. In this mode, the minerals constituted of heavy elements are lighter than minerals constituted of light elements. The minerals, which have the same gray color, have, more or less, the same chemical composition. For each sample, two photographs are presented at two different scales: on the left the scale bar is 50 microns and 10 microns in the right one. On the left photographs, the black part on the right of the picture corresponds to the outer part of the sherd. The black areas inside the sherd correspond to voids in the ceramics.
Figure 3A, B (opposite page): ET1 The minerals indicated as Fd corresponds to mantled feldspars, the core is made of albite whereas the outer part is made of K-feldspar. The crystal Q is quartz and the light associated zone is K-feldspar.

Figure 3C, D (opposite page): ET2 This sample is quite different from ET1. It presents numerous melted zones, which can be recognized by the bubbles included in a homogeneous matrix (arrows). The crystal on the left part of the left photograph corresponds to a mixing of two minerals. The lighter part is made of plagioclase, the gray one has a composition which do not corresponds to a definite mineral. Q = quartz

Figure 3E, F (page 44): ET3 This sample presents also evidence of melting indicated by fragments with bubbles, which probably corresponds to pockets where the melting collected. The melting processes are discrete around the minerals and less important than in ET2, This may be a consequence of a lower temperature of firing. Mantled feldspars like in ET1 are also seen (Fd).

Figure 3G, H (page 44): ET4. This sample is quite different from the previous ones and ET5 (see F, G). The mineral fragments are mainly quartz and phyllosilicates (elongated white crystals). The white minerals in the lower part of picture G are plagioclases (Pg), but K-feldspars are also found. The temperature of firing was probably lower than in the other samples.

Figure 3F, G (page 45): ET5. This sample shows melting evidences of the same type as ET3 and mantled feldspars. Some phyllosilicates are also present.

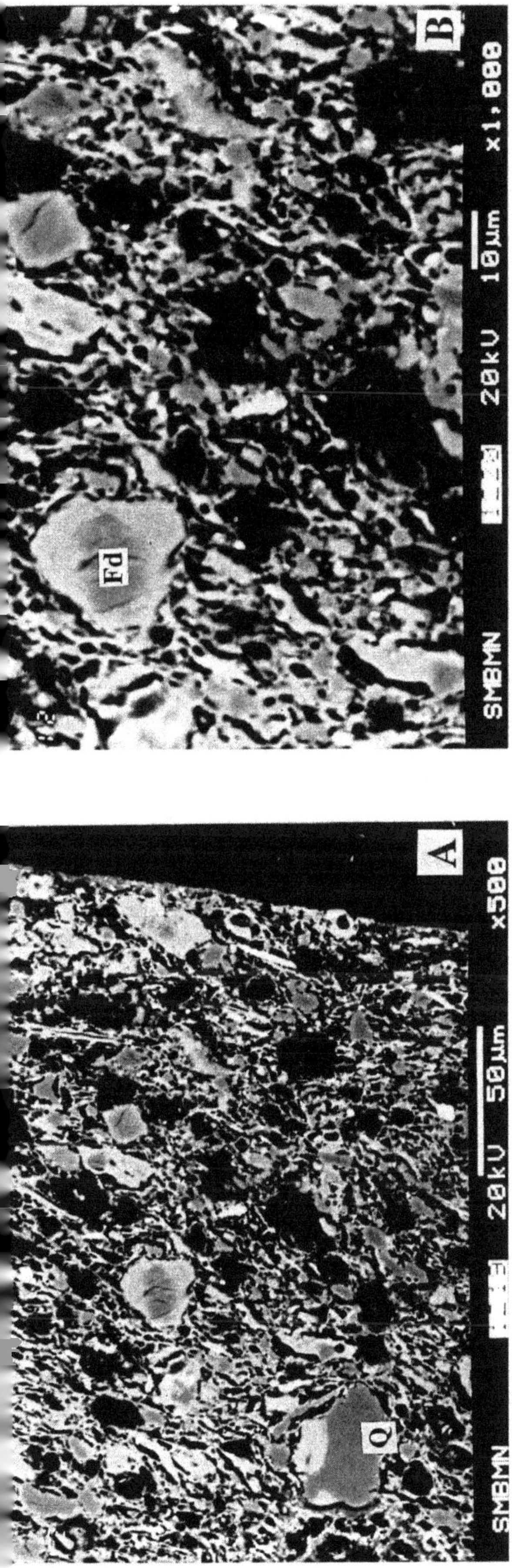
B
Fd
SMBMN 20kU 10µm x1,000
A
Q
SMBMN 20kU 50µm x500

ET 1

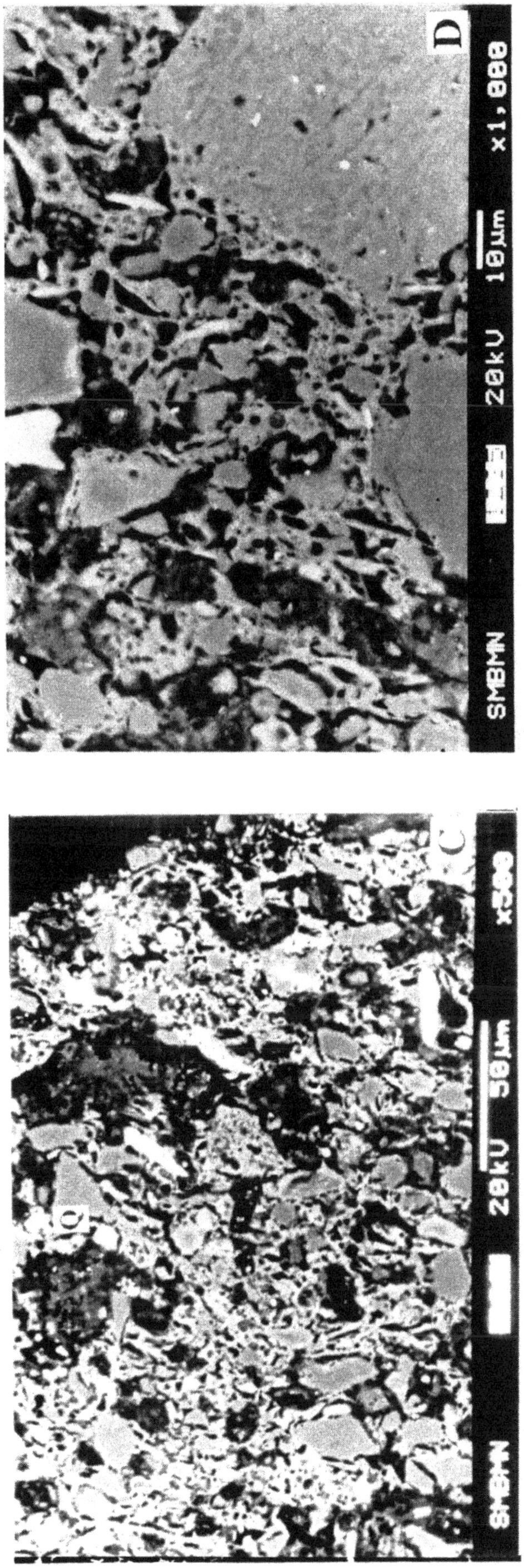
D
SMBMN 20kU 10µm x1,000
C
Q

ET 2

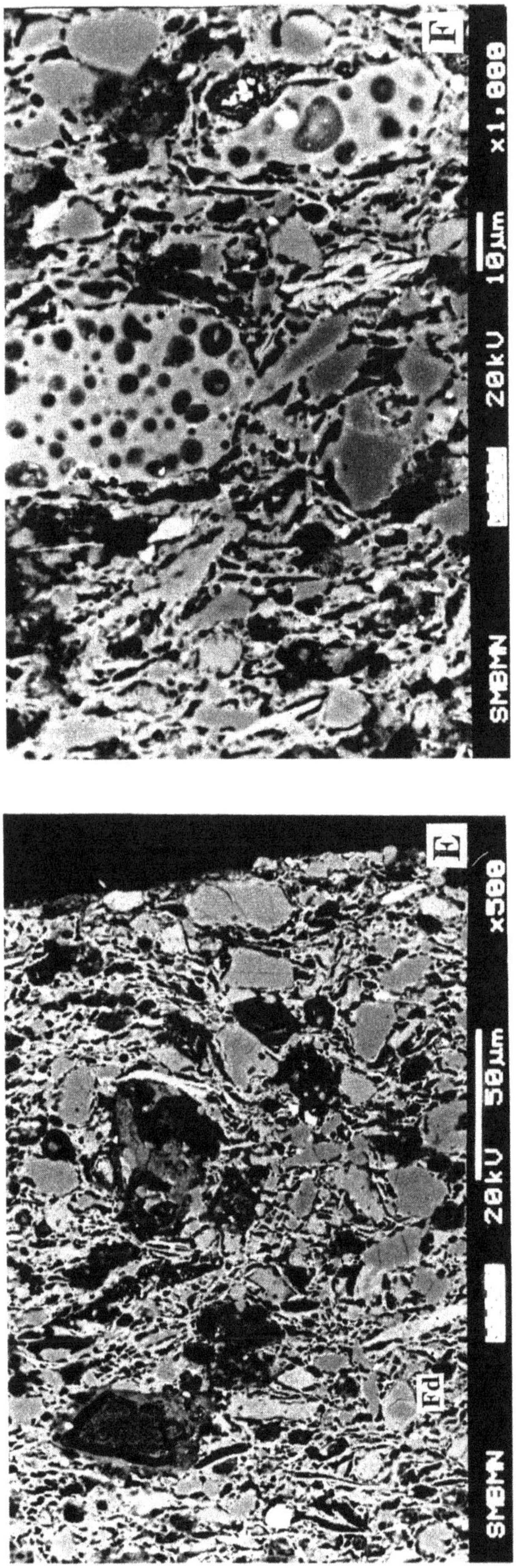
F
SMBMN 20kU 10µm x1,000
E
Fd
SMBMN 20kU 50µm x500

ET 3

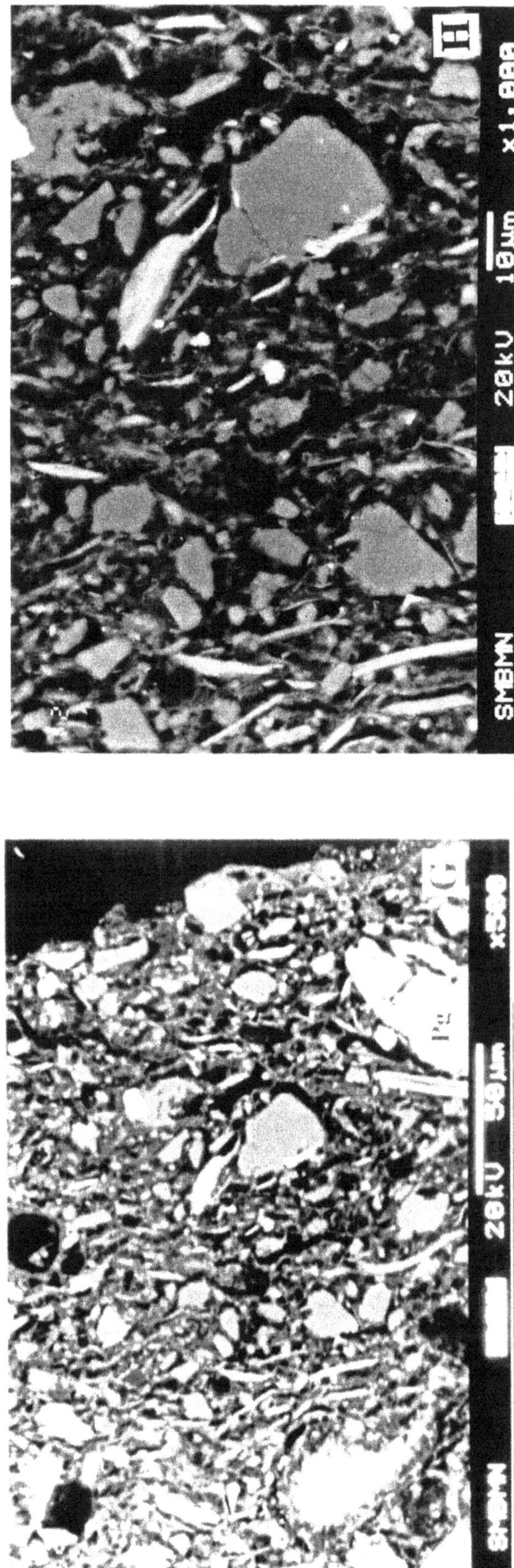
H
SMBMN 20kU 10µm x1,000
G
Pg
20kU 50µm x500

ET 4

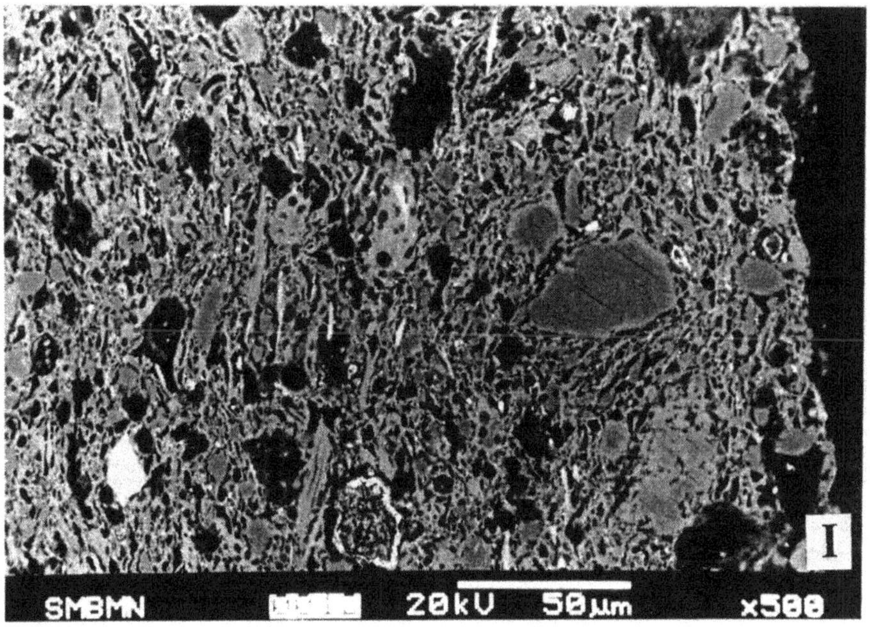

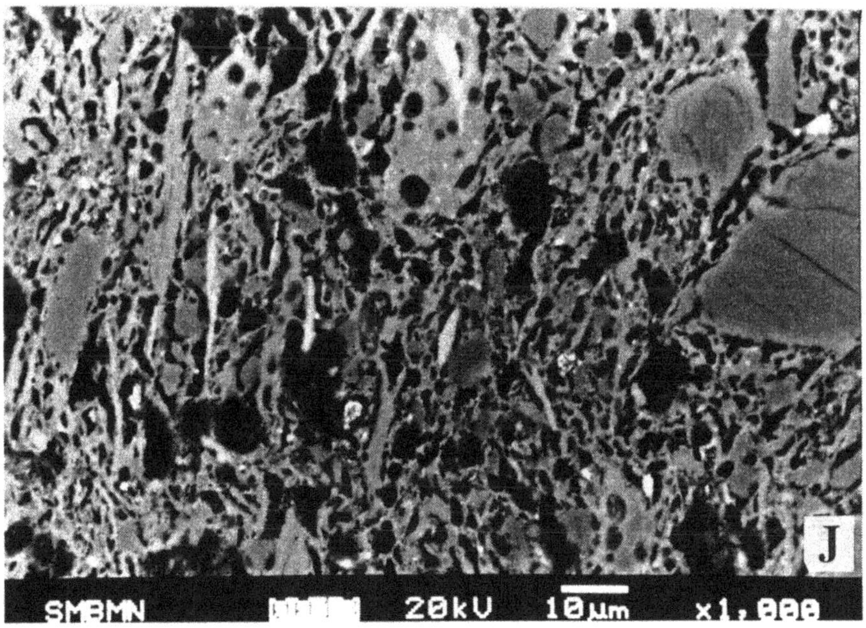

ET 5

ANALYTICAL RESULTS

In the five samples studied, 22 to 44 spot analysis of the matrix were performed by sample. The results of the analysis are presented as oxide weight percents (Table 2). In order to avoid too many tables, only the mean analysis is given in Table 2, but more detailed data may be obtained on request from the first author. It can be seen that ET4 has the highest value for silica and the lowest lime. The titanium oxide has not been determined in ET5, but in all the other analyses, its value is low with a high mean deviation.

Table 2. Arithmetical mean values and deviations for the SEM-EDS analysis of the matrix of samples ET1 to ET5. X=number of analyses.

SAMPLES	ET1		ET2		ET3		ET4		ET5	
Oxydes %	Mean	deviation	Mean	deviation	Mean	deviation	Mean	deviation	Mean	deviation
SiO2	53.47	3.48	52.89	5.91	50.47	0,04	61.6	5.73	57.77	8.48
TiO2	0.62	0.38	0.55	0.43	0.40	0.27	0.77	0.84		
Al2O3	21.68	3.10	24.05	4.02	26.32	3.57	22.74	3.87	21.21	4.54
FeO	5.72	1.99	6.26	2.49	5.27	2.82	6.17	2.16	4.86	2.98
MgO	3.11	1.15	4.06	4.09	3.56	1.92	1.59	0.70	2.74	1.31
CaO	9.10	3.24	6.50	4.04	7.82	3.59	4.30	1.30	7.69	3.20
Na2O	2.71	1.10	2.69	1.09	2.67	1.15	1.02	0.84	1.68	0.86
K2O	3.16	1.34	3.1	1.85	3.57	1.73	1.96	1.07	4.16	1.73
X	22		36		19		44		23	

In terms of relative weight percent SiO_2, Al_2O_3, and CaO of the calculated elements we found (except for ET4 where FeO is the third most important), then FeO (CaO in ET4) and MgO are the next most important. In this set of analyses, no P_2O_5 was found a result at variance with the analysis of the Nîmes sample, where the P_2O_5 content was interpreted as addition of bone powder. In fact, bone ashes are used to make porcelain in Britain and some other countries for the production of bone china (Schüller 1984). One may suppose that Etruscan people had discovered that P_2O_5 could produce ceramics of different quality in terms of their aesthetics and physical properties.

SiO_2-Al_2O_3-CaO diagram

Ternary diagrams help one to visualize the distribution of the samples according to their composition, considering only three components. The SiO_2-Al_2O_3-CaO diagram (figure 4, opposite page) takes into account the three most abundant oxides. It shows that the composition of the matrix is not very different from one sample to another. The areas of dispersion overlap considerably implying that all the original clays were similar. ET5 and the Nîmes sample are very similar, as well as ET1 and ET2. ET3 and ET4 are slightly different in their mean values, with low values in CaO and a greater variability of the Si/Al ratio.

Al_2O_3-MgO-FeO diagram

This diagram (Figure 5, opposite page) shows elongated areas of dispersion of the representative points with a nearly constant Fe/Mg ratio close to 0.66. However, it can be seen that there are characteristics in common (shape of the area, orientation of the main dimension), between samples ET1 and ET2, which are different from the other ones. ET5 and the Nîmes sample (sample 6) are similar but sample ET5 is slightly more aluminous. The value of the Fe/Mg ratio is probably a function of the parent-rock, which indicates that ET4, with a Fe/Mg ratio of 0.80, has a different parentage from the other sample. This is confirmed by Figure 3 G, H, where the structure and the composition of the ceramic ET4 can be seen to be different from the others. For example, the matrix is less oriented, the included minerals are mainly quartz and phyllosilicates, there is no fusion of minerals (cf. sample ET2) or mantled feldspars (cf. samples ET1 and ET5), nor inclusions containing bubbles like in samples ET3 and ET5. The simplest interpretation is that the original clay used by the potter was different from that of the other samples.

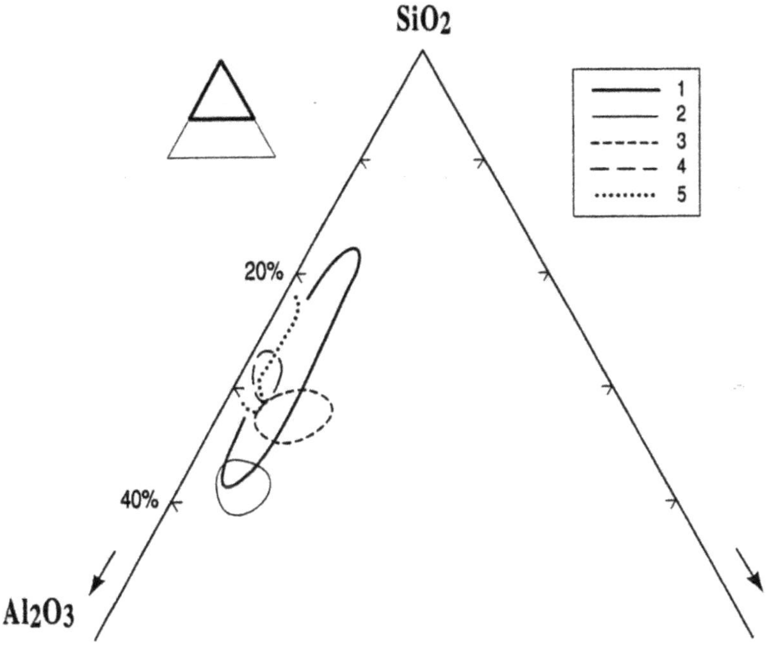

Figure 4. SiO_2-Al_2O_3-CaO diagram. Each number refers to the sample number, in this diagram the Nîmes sample is not plotted (see figure 3, in Bossière and Moguedet 1995).

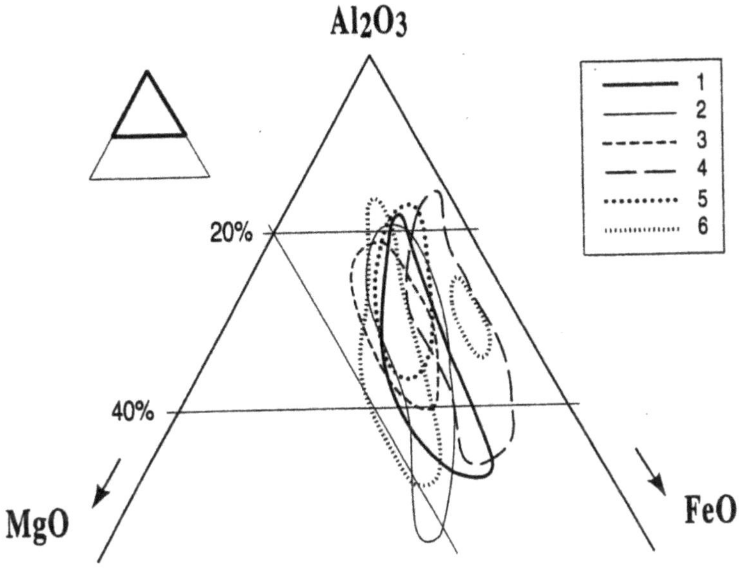

Figure 5. Al_2O_3-MgO-FeO diagram, each number refers to the number of sample; except 6, which refers to the sample from Nîmes.

CaO-Na_2O-K_2O diagram

This diagram indicates a greater variability (Figure 6), and the Na_2O/K_2O ratio varies from 0.5 to 0.79 with each ceramic having its own value in this range. This may suggest the addition of an alkaline fluxing agent, added in variable proportions by each potter. The variability may also be a consequence of the fact that it was difficult for the potter to control the added quantity of fluxing agent. Nevertheless, sample ET5 and the Nîmes sample (sample 6) are very similar. Sample ET4 covers a large domain. Samples ET1, ET2, and ET3 seem to have a common ratio between soda and potash.

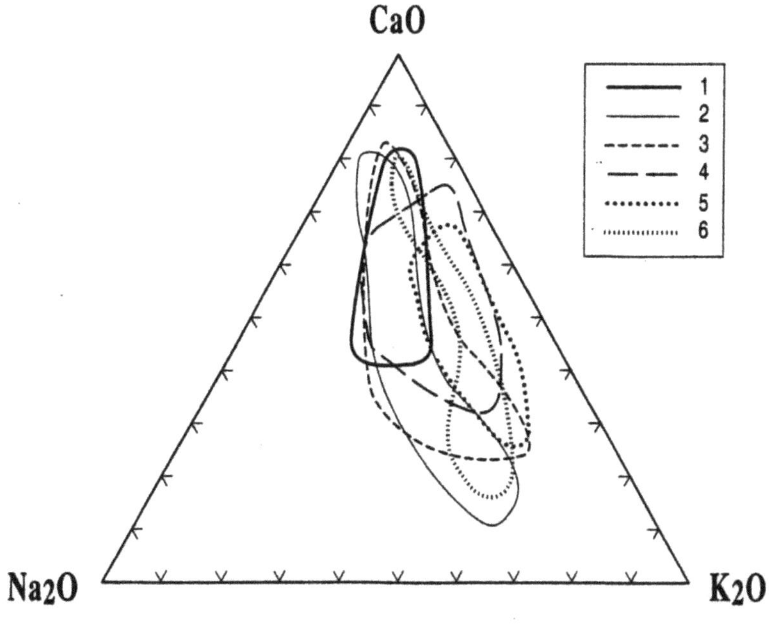

Figure 6. CaO-Na_2O-K_2O diagram. Each number refers to the sample number except 6, which refers to sample from Nîmes.

ACF diagram

This diagram (Figure 7) is commonly used for the study of the paragenesis of calcic metamorphic rocks with an excess of silica (free quartz). It allows also a determination of the parent rock. The parameters correspond to the values:

$$A = [Al_2O_3] - ([Na_2O] + [K_2O])$$
$$C = [CaO]$$
$$F = [FeO] + [MgO] + [MnO]$$

They are calculated by dividing the weight % of the analytical values by the molecular weight of the corresponding oxide.

In figure 7, the different representative areas have a more or less round shape, except for sample ET4, which has a slightly elliptic one corresponding to a greywacke composition. The long axis implies a more or less constant A/C ratio. The overlapping chemical compositions correspond to marl containing 35 to 65% of carbonate. The clay size fraction may originate from the alteration of volcanic or plutonic rocks of dacitic or granodioritic composition or any similar composition. In this diagram, sample ET5 and the Nîmes samples (sample 6) are very close, while ET1, ET2 and ET3 present similarities. Samples ET2 and ET6, which present relatively important partial melting show two areas of dispersion.

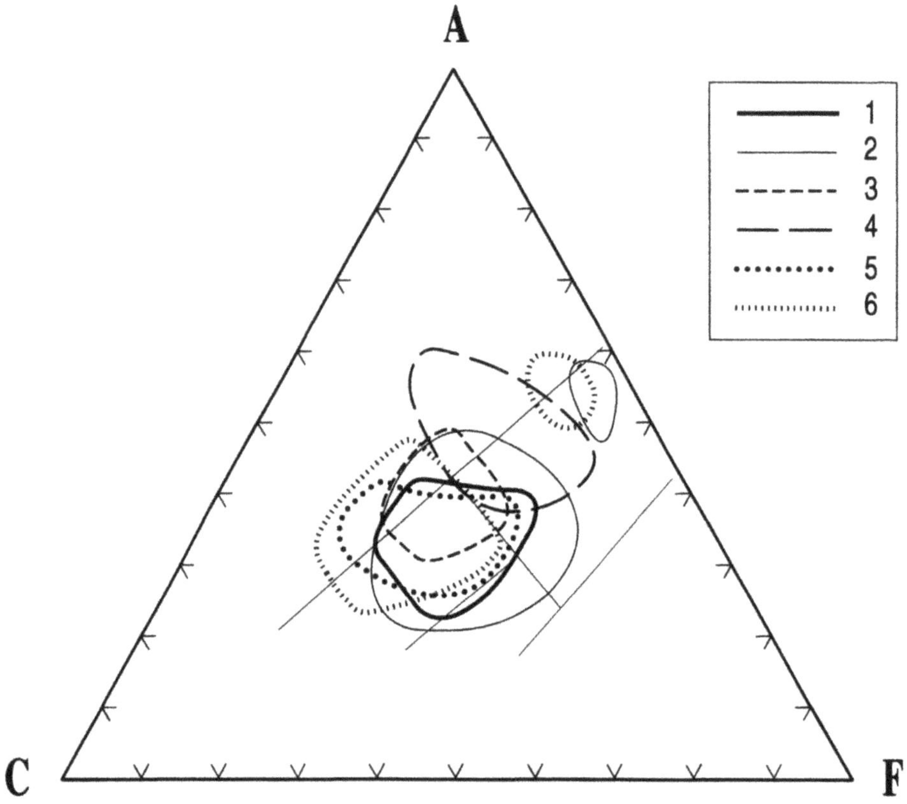

Figure 7. ACF diagram. The Al-rich clays and shales plot near the A corner. The surface limited by straight lines corresponds to clays and shales either free of carbonates or containing up to 35% carbonate. The higher C-component half corresponds to the composition of greywackes. The area out of this zone corresponds to marl containing 35-65 % carbonate. Most of the compositions in this area and in ET4 have a greywacke composition. Each number refers to the number of sample except ET6, which refers to sample from Nîmes.

Two samples have two representative areas: (1) Nimes (sample 6) represents as previously stated, a probable eutectic, and (2) ET2 shows melting around some minerals (Figure 3 C, D). These samples cannot belong to two different groups with and without carbonates. One must consider, therefore these spatial differences in composition to be a consequence of firing. The very restricted variability of the chemical composition of the lime free areas is interpreted as the result of a eutectic melting during firing. The chemical composition of each sample is different and, consequently, the chemical composition of the eutectic is different. An alternative interpretation might be that these products correspond to the growth of some new minerals, but the study in BSE do not show any difference of composition in these areas. They have a quite homogeneous color.

Sample ET4 again confirms its peculiarities. The shape of the dispersion suggests a greywacke origin and, consequently, a different parent rock, and, therefore, a different origin.

A comparison with the chemical composition of common plutonic rocks in an ACF diagram (Winkler 1979) shows that the arithmetical mean values (Table 2) of sample ET1 plots exactly on the mean value of tonalite (quartz andesite), sample ET5 plots exactly on the mean value of granodiorite (dacite). Samples ET2 and ET3 plots on each side of the mean composition of calc-alkali granites.

Chemical nature of the minerals in the matrix

Only the minerals grains set in the matrix are considered here. They may be different of the minerals picked up on the surface of the ceramics, and studied in the Nîmes sample by Bossière and Moguedet (1995). We consider here matrix minerals from samples ET1 to ET4 where the most frequent minerals being quartz, feldspars and phyllosilicates.

In all samples, quartz, as observed with the optical microscope and SEM, lacks any special shape. It generally occurs as more or less equidimensional grains with sharp edges and a mean diameter of 0.05 to 0.08 mm. Some crystals may be elongated (e.g. 0.08 * 0.03 mm).

The chemical EDS-SEM analysis of the phyllosilicates shows that they are neither a pure high temperature nor part of a biotite solid solutions series. They possibly represent mixtures resulting from weathering, previous to their use in the ceramics, or changes due to the firing of the clay. For example, in sample ET1 and ET3, the minerals have the Mg and Fe content of biotite, but the K value is not high enough. They are probably interlayered mixtures of biotite and chlorite. The phyllosilicates from samples ET1 and ET3 have nearly the same composition. In the sample ET3, there is also a white mica with the same composition as that in sample ET2. If one considers the distribution of Fe vs. Mg, and the nature of the phyllosilicates, there are at least two groups of ceramics.

Mantled feldspars were found in all the samples, except ET4. They are made of a core of plagioclase (albite) mantled by a K-feldspar with a variable Na content. The features shown in Figure 3 (samples ET3 and ET5) are quite characteristic. An unsolved question is: are these structures primary or a consequence of the firing? The variable content of the sodium could be an argument for a secondary origin due to firing.

Samples ET2 and ET5 also have pyroxenes (Table 3, Figure 8). Because muscovite does not naturally occur in (sample ET2) rocks containing pyroxenes, they must have been added (as part of the temper) to the original clay. From figure 8, it is clear that pyroxenes from ET2 and ET5 do not have exactly the same composition, the first one is a salite and the second one a diopside, nevertheless they show the same chemical evolution: Wo increases with the ferrosilite content. Such pyroxenes are present in rocks of alkali basalt parentage (Deer et al. 1997), where they present the same kind of chemical evolution. For example salite with an equivalent high-alumina content have been described in the Roman Volcanic Region (Cundari 1975). Consequently, the materials have different origins, and the ceramics do not have the same provenance. However, we need more data on pyroxenes from both sherds and geological samples to conclude on geographical localization.

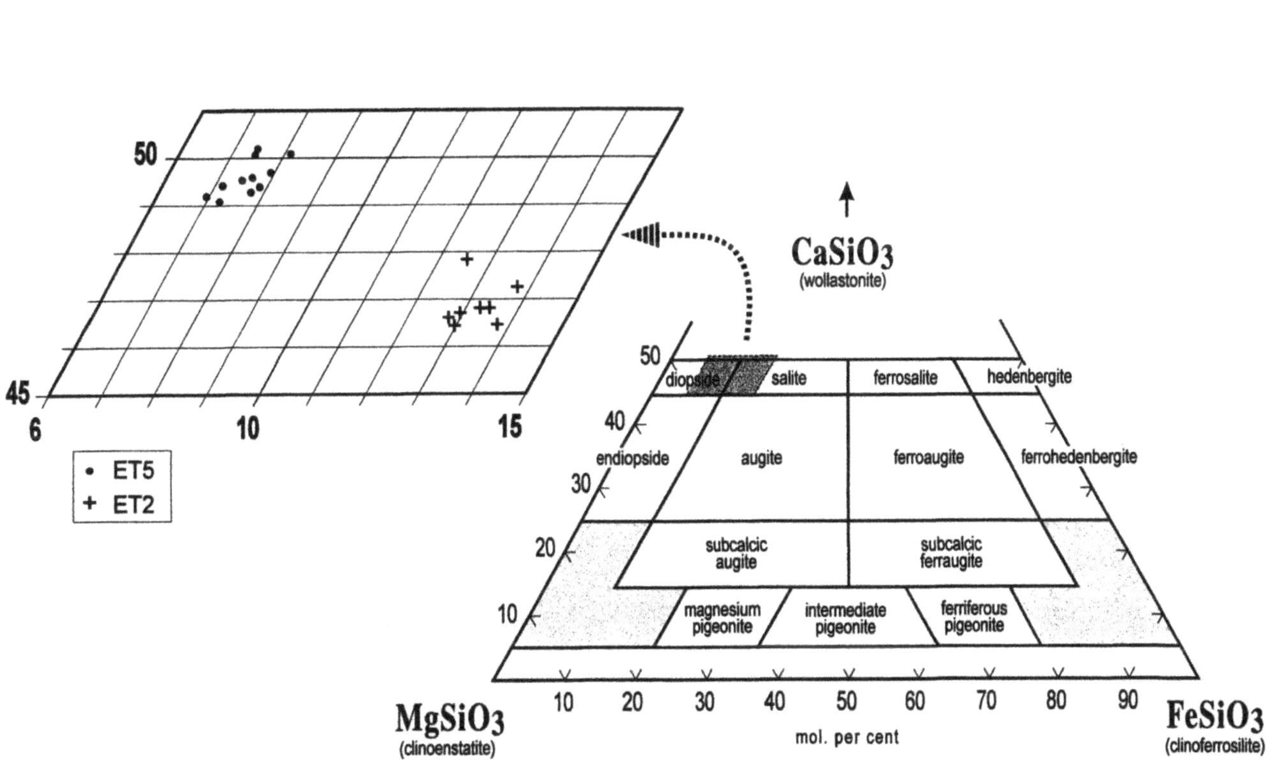

Figure 8. Chemical composition of pyroxenes from samples ET2 (crosses), and ET5 (dots), in terms of wollastonite (Wo), Enstatite (En) and Ferrosilite (Fe).

Table 3. Representative chemical compositions of clinopyroxenes from samples ET2 and ET5. #mg = Mg/(Mg+Fe).

	Sample ET2					Sample ET5				
anal n°	29	30	31	32	33	11	12	13	14	15
SiO2	49.84	48.84	49.76	49.44	48.17	50.72	51.47	50.55	51.4	51
Al2O3	4.44	4.31	4.01	4.16	4.29	2.89	2.97	2.66	2.82	2.87
TiO2	0.85	0.8	0.77	0.72	0.98	0.67	0.45	0.58	0.67	0.62
FeO	8.08	7.67	7.85	7.8	8.32	4.97	4.97	4.79	4.64	5.17
MnO	0.25	0.17	0.3	0.1	0.21	0	0.04	0.1	0.03	0.06
MgO	14.13	13.6	14.58	14.03	13.83	15.21	15.55	15.54	15.749	14.94
CaO	22.79	22.96	22.83	22.37	22.7	24.44	24.79	24.92	24.56	24.82
Na2O	0.39	0.3	0.31	0.42	0.34	0.08	0.12	0.12	0.09	0.2
K2O	0	0	0	0	0	0	0	0	0.1	0
sum	100.77	98.65	100.41	99.04	98.84	98.98	100.36	99.26	99.8	99.68
Si	1.85	1.852	1.853	1.863	1.831	1.894	1.895	1.886	1.896	1.895
Al	0.15	0.148	0.147	0.137	0.169	0.106	0.105	0.114	0.104	0.105
Al	0.044	0.044	0.029	0.047	0.023	0.021	0.023	0.003	0.018	0.02
Ti	0.023	0.022	0.021	0.02	0.028	0.018	0.012	0.016	0.018	0.017
Fe	0.25	0.243	0.244	0.245	0.264	0.155	0.153	0.149	0.143	0.16
Mn	0.007	0.005	0.009	0.003	0.006	0	0.001	0.003	0	0.001
Mg	0.781	0.768	0.809	0.787	0.783	0.846	0.853	0.864	0.865	0.827
Ca	0.906	0.932	0.911	0.903	0.924	0.978	0.978	0.996	0.97	0.988
Na	0.028	0.022	0.022	0.03	0.025	0.005	0.008	0.008	0.006	0.014
K	0	0	0	0	0	0	0	0	0.004	0
#mg	75	76	77	76	75	84	85	85	85	84
Wo	47	48	46	47	47	49	49	50	49	50
En	40	40	41	41	40	43	43	43	44	42
Fs	13	13	12	13	13	8	8	7	7	8

The chemical characteristics of the mineral grains set in the matrix yield information on the addition of temper and on the origin of the ceramics. This information comes from the chemical nature of a mineral itself or from the association of minerals. For example, pyroxene and muscovitic phyllosilicate cannot coexist in a same parent rock.

Melting occurrences

Two types of melting had occurred in the samples. The first one has been described for sample ET4, where melting is associated with carbon-rich globules. Plate 1 (middle of the volume) shows this kind of structure in a composite figure made of false-color compositional maps for Si, Al, Fe, Mg, Ca, Na, K and C. On the BSE image, black areas can be seen. On the element maps (except for the C map), the black areas have always a black core or a black corona. On the carbon map, carbon is light and two kinds of clear zones can be seen. Around the carbon-rich zones there is a melt corona with a more or less constant composition (Table 4). The outer one is lighter in BSE and the main oxides are silica, alumina, iron and lime. The sum is around 65 and the deficit from 100 corresponds to the C content. The darker zone in BSE has also a more or less constant composition but only for a globule in this case it is richer in carbon as the sum of oxides is around 40/43. This kind of melting has only been found in this sample, which is known to be of local production. A new examination of the thin sections has shown that melting associated to carbon-rich globules is also observed in samples ET2, ET3 and ET5. The C-rich melting products are still under study.

Table 4. Melted product compositions. The oxide sum of the carbon-rich products differ from 100 due to the carbon content.

Sample ET4	Globule 1						Globule 2					
	Outer zone			Inner zone			Outer zone			Inner zone		
N° anal.	10	11	12	18	19	20	38	39	43	34	35	36
SiO2	46.61	47.15	47.73	20.64	20.02	17.78	42.53	30.95	48.92	26.98	27.42	27.21
Al2O3	9.84	9.09	9.33	11.83	11.94	9.9	9.77	10.3	11.64	8.9	9.35	9.75
TiO2	0.2	0.15	0.23	0.1	0.08	0.27	0.22	0.17	0.5	0	0.07	0
FeO	1.88	1.38	3.56	3.27	3.13	4.35	2.38	3.04	1.97	0.91	0.98	1.4
MnO	0.08	0	0	1.63	1.86	2.31	0	0.09	0.08	0	0.06	0.04
MgO	0.91	0.78	1.38	0.32	0.27	0.33	1.67	2.02	1.36	0.63	0.46	0.48
CaO	3.09	3.16	3.4	2.29	2.34	1.89	2.57	2.04	3.47	1.68	1.64	1.75
Na2O	0.11	0.11	0	0	0	0	0.16	0.13	0.11	0.05	0.11	0.04
K2O	0.66	0.72	0.78	0.33	0.18	0.23	0.66	0.93	0.7	0.37	0.36	0.45
SO2	tr	tr	tr	0.87	1.02	0.82	0.12	0.22	0.22	0.17	0.27	0.07
P2O5	tr	tr	tr	1.92	2.2	1.7	0.69	0.48	0.84	0.89	0.92	0.69
Sum	63.38	62.54	66.41	43.2	43.04	39.58	60.77	50.37	69.84	40.58	41.64	41.88

The second type of melting corresponds to the melting of minerals. It has been found in samples ET2, ET3 and ET5. Figure 9 shows an example of melting of a biotite crystal in sample ET2.

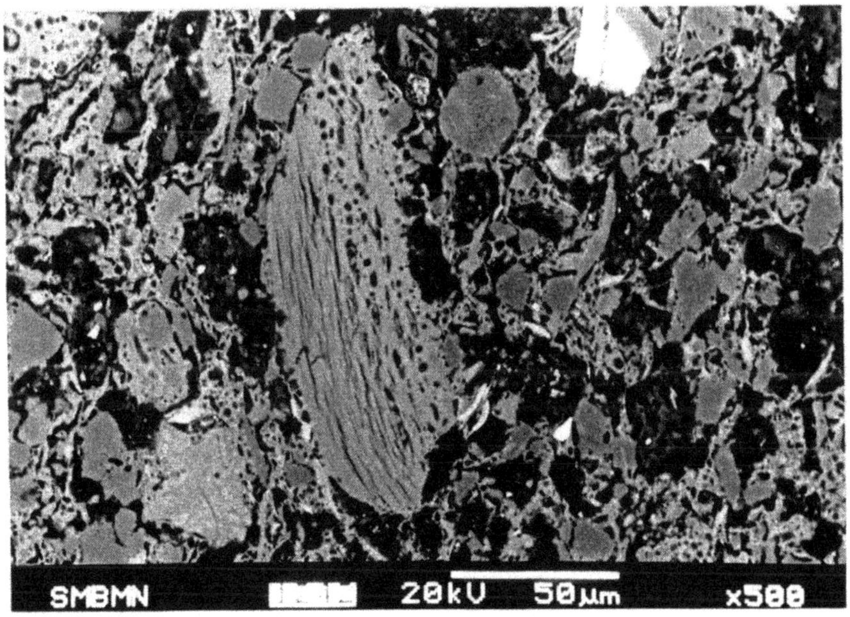

Figure 9. Melting of a crystal of biotite.

DISCUSSION AND CONCLUSION

The mineralogical/chemical modifications of the ceramics that take place during firing should be studied in order to acquire a better knowledge of both the origins of the clay and the firing process. It is clear that in some cases melting of minerals occur and not in others (Figure 3). The question is why? Is it linked to the chemical composition of the original clay, a high temperature of firing, or the addition of a fluxing agent? For a given mineralogical-chemical composition, eutectic melting occurs at a given temperature. This temperature differs according to the clay composition and the nature of the fluxing agents. In the case of sample ET4, melting is linked to carbon-rich globules. Were these globules present in the original clay? Or did the potters add carbon-rich material to their pots? The question is still unsolved, but it is known that the so-called Bucchero Nero ceramics are so rich in carbon that they don't need to be carbon coated to be studied under a SEM because they are already electric conductors.

In terms of ceramic provenance and origin of the clay material, all the diagrams and BSE images and included minerals show that sample ET4 is different. This suggests that the clay of sample ET4 differs in origin and parent-rock from the other samples. Furthermore, the analyses show that it has a composition of a greywacke. ET4 is known by its archaeological context to be a local production, and it is well characterized by its chemical and mineralogical composition. The question is then why the other samples are different? They may be imported and there are no major objections from the archaeologist's point of view against such a hypothesis. They also could have been made out of a different material, but still local, only geological field works, not already done, may give an answer to this question. The problem is that only one local sample, to date has been studied. The origin of ET2 was questionable from the archaeologist's criteria (Table 1), the similarities of its chemical composition with the other samples allow to conclude that it is closer to samples from southern Etruria than from Inland Campania.

It is also clear from the nature of the pyroxenes that sample ET2 and ET5 have different origins because the included pyroxenes are different. But ET5 and the Nimes sample ET6 are very similar and probably from the same material, as ET1 and ET2 form another similar pair. All these samples are different from ET4 from every point of view, and have, from the ACF diagram (figure 7), a composition influenced by a marly component. They may be derived from the weathering of dacitic rocks, occurring in the area, or from imported rocks of equivalent composition and they were probably imported because they are different from every point of view of ET4.

In summary, ET4 is local and made out of a greywacke-clay (a clay with the composition of a greywacke), ET2 and ET5 have an added temper originated from rocks of alkali basalt parentage but from different origin. From the chemical composition of the matrix ET1, ET2, and ET3 have similarities as well as ET5 and the Nimes sample ET6. They have a similar, dacitic-rock origin. ET2, ET3, and ET5 have melting in common but ET3 have not an added temper.

An important result of this study is the evidence of melting which may occur as a eutectic (samples ET2 and Nîmes sample ET6), a mineral fusion (ET2), or as the result of melting around carbon-rich inclusions (ET4, ET2, ET3 and ET5). These characteristics should be of major interest for the determination of the origin of the sherds, but the study is at its beginning.

In conclusion, in order to determine the origin of a sherd it is helpful:

(1) To look for traces of melting.
(2) To study the chemical nature of the minerals grains included in the matrix because their chemical characteristics are (a), useful indicators of the differences in the origin of the clays or soils used in ceramic manufacture, or (b), indicate the addition of temper.
(3) To study the chemical nature of the matrix because it may be complex.

In the future, it would be interesting to obtain more detailed data for two areas: south Etruria and the inland Campania. It would also be interesting to extend the study to different ceramic types like Bucchero Nero and amphora.

Acknowledgments

We gratefully acknowledge: M. Stefano De Caro, superintendent of Campania who allowed these analysis; Mrs. Claude Albore-Livadie, research manager (French CNRS), who is at the origin of this work, A. Barreau, electronics engineer in charge of the SEM and D. Shelley (University of Canterbury, Christchurch, N. Z.) who kindly reviewed the English.

References

Albore Livadie, C. 1979. Notes de typologie et de chronologie. In *Le bucchero nero étrusque et sa diffusion en Gaule*, Actes de la table-ronde d'Aix-en-Provence. Bruxelles: Latomus, pp. 91-130.

Amyx, D.A. 1988. *Corinthian Vase-Painting of the Archaic Period.* Berkeley and Los-Angeles: University of California Press.

Belleli, V. 1997. Dal Museo di Tarquinia: decoratori etruschi di "Running Dogs", in *Miscellanea etrusco-italica. II*, Roma: Consiglio Nazionale delle Ricerche, pp. 7-54

Boardman, J. 1998. *Early Greek Vase Painting 11th-6th centuries BC. A handbook.* London: Thames and Hudson.

Bossière, G., and Moguedet, G. 1995. Etude chimico-minéralogique de céramiques étrusco-corinthiennes d'Etrurie méridionale: contribution à la connaissance de l'origine du matériel et des techniques employées. *Revue des Etudes Anciennes* 97(1-2): 5-26.

Colonna, G. 1961. La ceramica etrusco-corinzia e la problematica storica dell'orientalizzante recente in Etruria. *Archeologia Classica* 13: 9-25.

Colonna, G. 1964. Aspetti e problemi della Roma primitiva: il periodo orientalizzante recente, *Archeologia Classica* 16: 1-32.

Cundari, A. 1975. Mineral Chemistry and Petrogenetic Aspects of the Vico Lavas, Roman Volcanic Region, Italy. *Contribution to Mineralogy and Petrology* 53: 129-144.

Deer, W. A., Howie, R. A., and Zussman, J. 1997. *Rock-Forming minerals, Vol. 2A Single-Chain Silicates.* Second edition. London: The Geological Society.

Druc, I., 2000. Soil sources for ceramic production in the Andes. *37th Annual Meeting The Clay Minerals Society, Chicago, June 24-29, 2000.*

Frère, D. 1995. Méthode pour l'étude des balsamaires italo-corinthiens à décor subgéométrique. *Revue des Etudes Anciennes* 97(1-2): 27-43.

Frère, D. 1995. *Recherches sur les aryballes et les alabastres italo-corinthiens à décor subgéométrique.* Doctoral Thesis. Nantes: Département d'histoire, Université de Nantes.

Meyer, J.C. 1993. *Pre-Republican Rome. An analysis of the cultural and chronological relations. 1000-500 B.C.*, Analecta Romana Instituti Danici, Suppl.11. Odense: Odense University Press.

Neeft, C.W. 1987. *Protocorinthian Subgeometric Aryballoi,* Allard Pierson Series 7. Amsterdam: Allard Pierson Museum.

Payne, H. 1931. *Necrocorinthia. A Study of Corinthian Art in the Archaic Period*, Oxford.

Schüller, K., H. 1984. Silicate Ceramics. In *Process Mineralogy of Ceramic Materials*. eds. W. Baumgart, A.C. Dunham, and G.C. Amstutz. Stuttgart: F. Enke Publishers. pp. 1-27

Spivey, N. J. 1988. *Un artista etrusco er il suo mondo. Il pittore di Micoli.* Studi di archeologia pubblicati dalla soprintendenza archeologica per l'Etruria meridionale 5. Roma: De Luca Editore.

Shriner, C. M., and Murray, H. H., 2000. Ceramic technology at Lerna, Greece in the third millenium B.C. *37th Annual Meeting The Clay Minerals Society, Chicago, June 24-29, 2000.*

Szilàgy, J.G. 1992. *Ceramica etrusco-corinzia figurata. Parte I. 625-590/580.* Monumenti etruschi 7. Firenze: L.S. Olschi.

Szilàgyi, J.G. 1998. Ceramica etrusco-corinzia figurata. Parte II. 590/580-550. Monumenti etruschi 7. Firenze: L.S. Olschi.

Velde, B. and Druc, I. 1999. *Archaeological ceramic materials: Origin and utilization.* Berlin: Springer-Verlag.

Winkler, H.G.F. 1979. *Petrogenesis of Metamorphic Rocks.* Fifth edition. Berlin: Springer-Verlag.

CHAPTER 5

CLAYS AND EARLY NEOLITHIC POTTERS[1]

Guilmine Eygun
Université de Montréal, Département d'anthropologie

In recent years, considerable efforts have been devoted to characterize the technological features of prehistoric ceramics and their direct relationship with the presumed function of pots. The results obtained from the petrographical analysis and X-ray diffraction performed on two Early Neolithic ceramic assemblages from South-eastern Italy, has evidenced an important technological variation suggesting that potters choices can be explained equally in terms of culturally learned behavior and functional strategies.

INTRODUCTION

With the increased sophistication in the laboratory analyses of archaeological vessels, the technological study of pottery has become one of the most common ways to approach prehistoric ceramics. It has undoubtedly led to a more detailed knowledge of production sequences as well as a more complete understanding of the socio-economic aspects of pottery. One consequence of this situation, however, is that many technological studies of ceramics have focused and continue to focus their interests on the functional aspects of pots in such a way that the emphasis is being put on the causal link between the use of a container and its technological properties. The underlying assumption is that there must be a direct correlation between the function of a pot and its physical properties (Braun 1983; Steponaitis 1983; Bronitsky 1986; Schiffer and Skibo 1987; O'Brien et al. 1994). Although technological variability encountered in ceramic vessels may be partly due to their intended use, different options can fulfil the same functional requirements. If environmental and functional constraints play an important role in pottery making, social traditions and individual arbitrary decisions can be as significant (Van der Leeuw 1993).

One example of this technological flexibility is illustrated by the results of petrographical analysis and X-ray diffraction performed on two Early Neolithic ceramic assemblages from Southeastern Italy. The present paper will attempt to demonstrate how the technological variations encountered may depend as much on socio-cultural constraints and personal choices as on presumed functional needs.

ARCHAEOLOGICAL AND GEOLOGICAL CONTEXTS

The ceramic assemblages come from the Early Neolithic sites of Ripa Tetta and Lagnano da Piede, both dating from the end of the fifth millennium BC (Ripa Tetta: 6890 ± 60BP, Lagnano da Piede: 6790 ± 255 BP and 6700 ± 100 BP). These sites are located in South-eastern Italy on the plain of the Tavoliere (Figure 1) which is one of the best known regions of Italy for the Neolithic period because of the exceptional density of ditched villages. The combined effects of extensive agriculture and limestone bedrock have permitted the identification of over 1,000 Neolithic sites through aerial photography (Bradford and Hunt 1946; Bradford 1949). As of today, between 80 and 90 sites have been archaeologically investigated spanning the Early to Late Neolithic. Contrary to the situation in most of the western Mediterranean where the neolithization process seems to appear progressively (Guilaine 1980; Barker 1981, 1985; Chapman and Müller 1990), Early Neolithic sites of the Tavoliere -and of south-eastern Italy in general- are characterized by a fully developed Neolithic mode of life with autonomous agro-pastoral economy, well established village settlement and technically developed ceramic production.

Because of their abundance and high visibility, ceramic complexes represent the most studied material of this period and play a fundamental role in the chronological framework of the area. Fine and coarse wares, impressed and/or painted, are encountered together in the earliest sites of the Neolithic of the Tavoliere although a controversy still exists about the presence of only one early coarse ware type of *ceramica impressa*. Little is known however about the practical use and the socio-cultural function of this pottery. Only recently several studies have attempted to analyze ceramics in this perspective thus opening new fields of investigation and understanding (Whitehouse 1991, 1992; Cassano 1993; Muntoni 1995; Eygun 1998, 1999).

[1] This chapter is a revised and expanded version of an article published in French by the author (Eygun 1999).

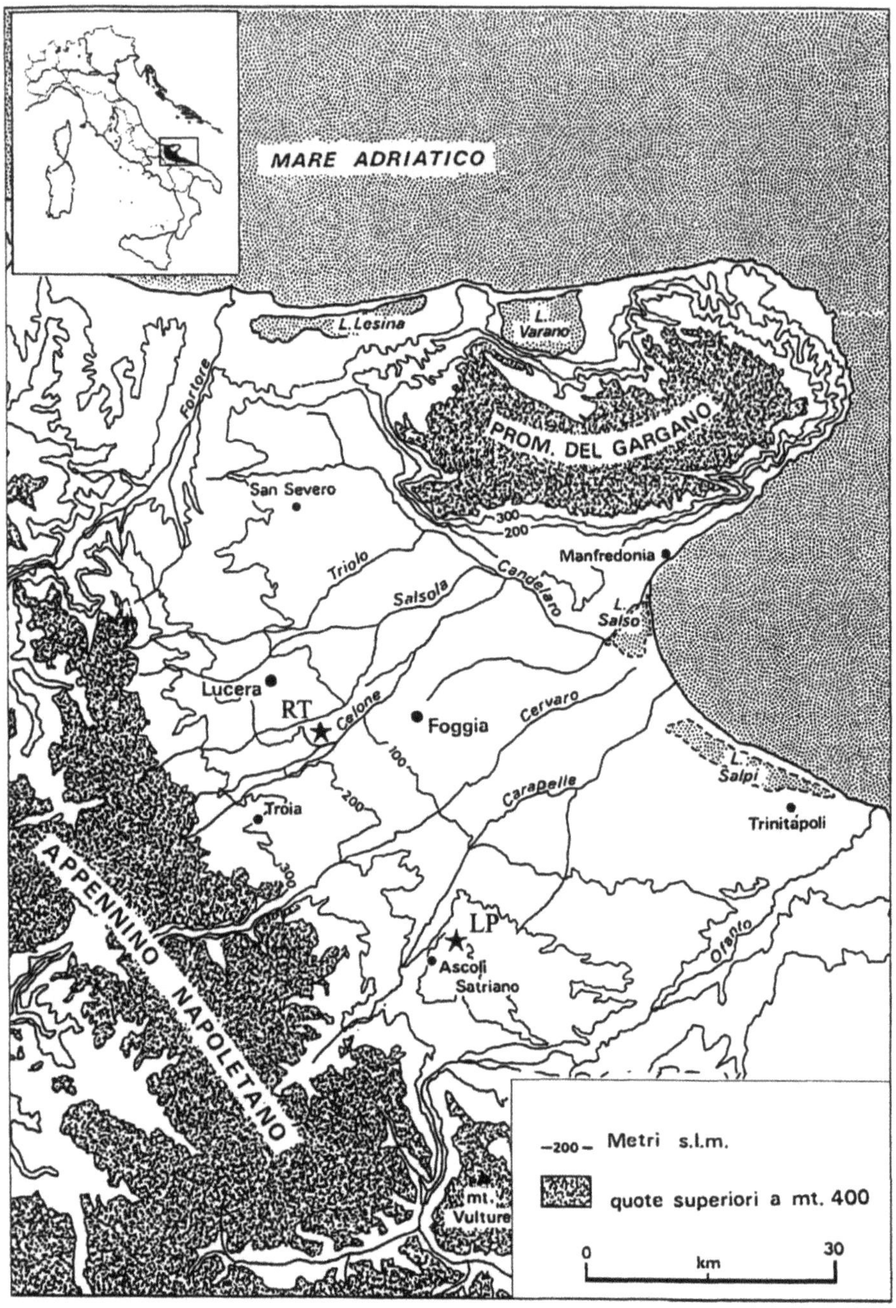

Figure 1. Map showing the location of the two Neolithic sites Ripa Tetta (RT) and Lagnano da Piede (LP) in the Tavoliere plain (after Eygun 1999, fig. 2)

Geologically, the Tavoliere plain is a limestone depression filled with marine deposits of Pliocene and Pleistocene clay, often overlain by later Pleistocene material and Holocene alluvial deposits. The geological composition of the soil is uniform for most of the plain with fine sub-Apennine clays sharing a very similar chimico-mineralogical composition regardless of their age and place formation (Dell'Anna et al. 1974; Balenzano et al. 1984; Cianco et al. 1995). The depth of the outcrops may vary from a few meters to 350 meters with an average thickness of 300-400 meters (Balenzano et al. 1977). Because of their mainly calcareous composition (may contain as much as 30% limestone), clays from the Tavoliere are considered today good materials for brick and tile production. This type of clay is also suitable for low-fired pottery (under 800-850°C) such as most Neolithic ceramics (Dell'Anna 1986; Dondi et al. 1992).

Both sites, Ripa Tetta located in the northwest part of the plain and Lagnano da Piede located 40 km away in the southwest part (Figure 1), are situated on major clay outcrops, which are still exploited today for brick production. Although other outcrops exist in the vicinity, the geological similarity of the clays from the Tavoliere strongly suggests that the most readily and closely available clay sources were the ones used by prehistoric potters. Access to water was also direct via a small river running few hundred meters away from each site. Basic components needed for ceramic production were thus present at both sites.

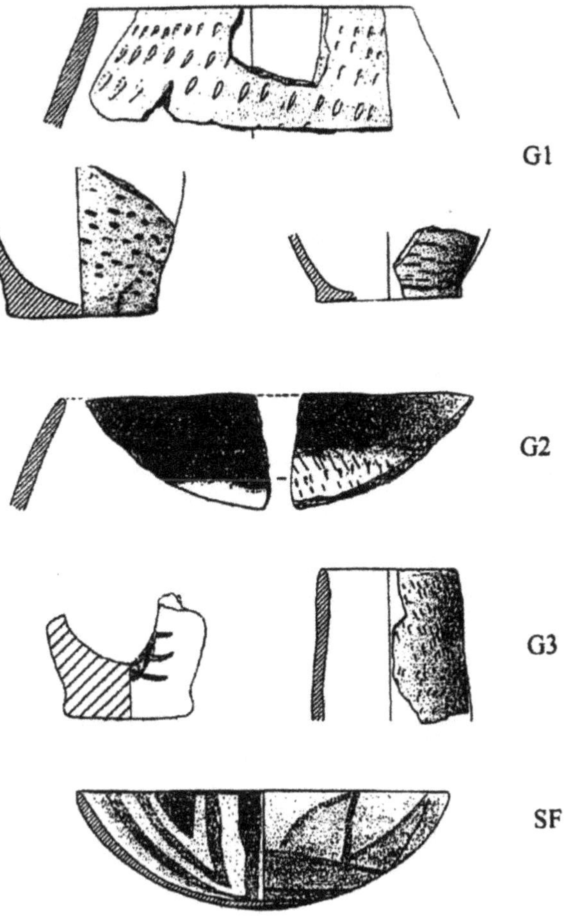

Figure 2. Four types of wares from Ripa Tetta and Lagnano da Piede (after Eygun 1999, fig. 3)

POTS AND PASTES

Four specific ware types have been identified from our ceramic assemblages on the basis of their technological, morphological and stylistic variation: three rather coarse wares termed *Grossolana* 1 (G1), *Grossolana* 2 (G2) and *Grossolana* 3 (G3), and one finer ware referred to as *semi-fine* (SF) (Figure 2). Coiling technique was the only construction technique observed on the body fragments of all vessel types from both sites; suggesting that all pots were made using this building procedure. Small pots may have also been made with the pinching technique although no traces have been identified. Frequency of building traces however is proportional to the size of the vases: the larger the vase, the more frequent the observed fracture between coils.

In both sites frequency of coarse wares is more important than that of fine wares; and within these main categories, frequency of each vessel type is very similar from one site to the other with slightly more fine ware in Lagnano da Piede and more frequent G1 and G2 vessels in Ripa Tetta (Fig. 3).

Coarse ware containers have a straight sided to closed shape with thick walls and flat bases. While G1 pots are characterized by an important size and form variation, most of the G2 vases are large to very large containers. On the other hand, G3 pots, which are mostly small to medium size, have a thick foot-like base attached to the rest of the vase. Fine ware containers, which are usually, have an open shape with round bases and thin walls. All categories are decorated. Impression and incision are primarily used on coarse ware, although many G2 containers also have a red slip applied as a large band on the upper part of both the interior and exterior walls. Fine ware is painted or exhibits both painting and impressions or incisions (Figure 2).

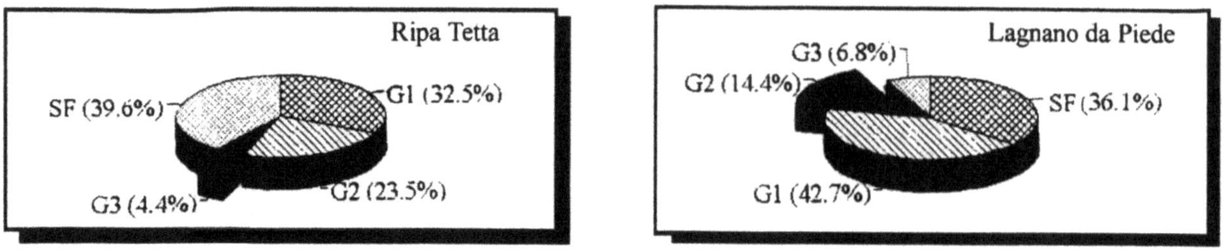

Figure 3. Type Frequencies

The mineral composition of ceramic samples from each group as well as from the clay outcrops from the two sites was determined both by petrographical analysis and X-ray diffraction. Seventy-eight fragments from the four ware types (fifty-three for Ripa Tetta and twenty-five for Lagnano da Piede) and eight samples from the local clays and conglomerate have been comparatively studied under a petrographical microscope.

Twenty-five samples from both sites representative of each ware type and five samples of local clays were then submitted to X-ray diffraction. Both qualitative (composition of the clay and non-clays minerals) and quantitative (mineral frequency, grain-size and grain-angularity distribution) analyses were performed on the samples. The complete results of the analysis, which are presented elsewhere (Eygun 1998), are summarized in Table 1.

Despite differences in color, and sometimes in texture, the clays are remarkably uniform in their mineral composition. All the minerals in the prehistoric fragments correspond to those of the local clays of Ripa Tetta and Lagnano da Piede (Figure 4). Consistently present are the clay mineral illite and dolomite with some traces of chlorite and montmorillonite, along with the non-clay minerals calcite, quartz, feldspar, pyroxene, volcanic rock and muscovite.

Volcanic rock fragments are naturally present in the clays and come from the Monte Vulture located about 60 km at the south of the Tavoliere. Volcanic rock could not be identified in the X-ray diffraction because it is an amorphous mineral and this method is based on the identification of the mineral structure. Presence of organic inclusions of marine and land shells as well as microfossils in all samples is mainly due to the marine origin of the clays from the Tavoliere. These inclusions are very frequent and are always naturally found in local clays and prehistoric wares. Kaolinite however is only found in the local raw clay samples which can be explained by the decomposition of the mineral at temperatures between 550°C and 625°C (Grim 1968) assuming that kaolinite was originally contained in the ceramic paste. Such an assumption is justified by the fact that, when the same raw clay samples are heated at 600°C, no traces of kaolinite are observed (Figure 5).

The exploitation of the local clay is not surprising as we know from the ethnoarchaeological record that in most situations of domestic production, the principle of least effort predominates among potters (Arnold 1985; Gosselain 1995).

While the clay used to make each of these wares is the same and most likely comes from the immediate clay outcrops, clearly indicating no environmental constraint, the petrographical study of the mineral composition of each group also shows that in most cases temper was added to the clays. Petrographical analyses indicate that these inclusions come from the conglomerate located stratigraphically under the clay beds. The types of inclusions present in the sand, which composes this conglomerate, however, are the same as those already present in the natural clays that is mainly quartz, feldspars, calcite and chert. They only differ in form and size. Added inclusions are coarser like those present in the conglomerate and, in the case of quartz, they can also be sub-rounded while they always have a subangular to angular shape in the natural clays (Table 1). On the other hand, presence of these inclusions tend to be dominant in the Neolithic wares while they are only present in small quantity in the natural clays suggesting that more inclusions of the same type were added. Therefore in this case, it is not the nature of the temper which varies, but its size, shape and quantity. Although this procedure is not commonly encountered in prehistoric contexts, it is known for more recent ceramic production (Velde and Druc 1999: 80-82).

If the types of inclusions present in the temper are the same as those naturally present in the clays, why would then the potters have added them to their pastes? In other words, what are the advantages of including in the clays the same tempering material as the one already present? The reasons for such choices may depend on two levels of decision-making: one related to the building procedures, and the other to the intended use of the pots.

Table 1. Mineralogical composition of the local clays and conglomerate and the four neolithic paste groups from Ripa Tetta and Lagnano da Piede (dark cells indicate major observed differences).

	LOCAL CLAY	CONGLO-MERATE	G1	G2	G3	SF
CLAY MINERALS						
KAOLINITE	P	/	ND	ND	ND	ND
ILLITE	P	/	P	P	P	P
CHLORITE	TR	/	TR	TR	TR	TR
DOLOMITE	P	/	P	P	P	P
MONTMORILLONITE	TR	/	TR	TR	TR	TR
NON-PLASTIC MINERALS						
QUARZT	C-D SI-M A-SA	D-VD SI-VL A-SR	C-D SI-L A-SR	D SI-L A-SR	D-VD SI-VL A-SR	C-D SI-M A-SR
CHERT	U-C SI-M A-SA	D-VD SI-VL A-SR	U-C SI-M A-SR	U-C SI-M A-SR	C-D SI-VL A-SR	U-C SI-F A-SR
CALCITE	C-D SI-M A-SR	D-VD SI-VL A-SR	FR-VD SI-L A-SR	FR-VD SI-L A-SR	U-C SI-VL SA-SR	C-D SI-L A-SR
FELDSPAR	VU-C SI-F A-SR	FR SI-VL A-SR	VU-C SI-M A-SR	VU-C SI-M A-SR	U-C SI-L A-SR	VU-FR SI-F A-SR
PYROXENE	VU-U VF-F A-SR	U SI-M A-SR	VU-U SI-M A-SR	VU-U SI-M A-SR	VU-U SI-M A-SR	VU-U SI-F A-SR
VOLCANIC ROCK	VU VF-F SR-R	U VF-F SR-R	ND-U VF-F SA-R	ND-U VF-F SA-R	ND-VU VF-F SR-R	ND-U VF-F SR-R
MUSCOVITE	VU-U SI-F A-SA	ND	VU-U VF-F A-SA	VU-U VF-F A-SA	ND-U VF A-SA	VU-U SI-F A-SA
ORGANIC INCLUSIONS						
SHELLS	U-FR	U-FR	ND-FR	ND-FR	ND-U	ND-FR
FOSSILS	U-FR	U-FR	ND-C	ND-C	ND-U	ND-C

FREQUENCY

VD	Very dominant
D	Dominant
C	Common
FR	Frequent
U	Uncommon
VU	Very uncommon
ND	Non detected
P	Present
TR	Traces

SIZE

VL	Very large (1000-2000μ)
L	Large (500-1000μ)
M	Medium (250-500μ)
F	Fine (125-250μ)
VF	Very fine (62-125μ)
SI	Silt(<62 μ)

SHAPE

A	Angular
SA	Subangular
SR	Subrounded
R	Rounded

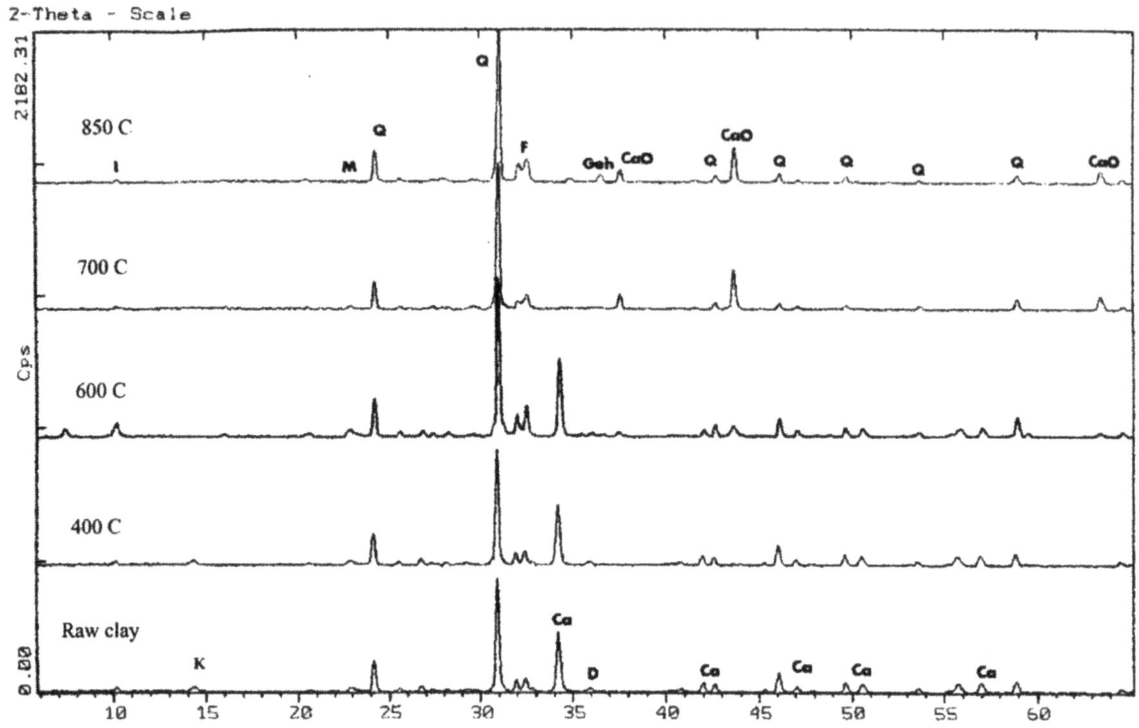

Figure 4. Comparative X-ray diffractions of local clay and neolithic wares. Their mineral composition is very similar. A new mineral, Gehlenite (Geh), appeared in the local clay sample as a result of its firing at 850°C. Ca = calcite, CaO=Calcium oxide, D= Dolomite, Geh= Gehlenite, F= Feldspar, I= Illite, M= Muscovite, Q= Quartz.

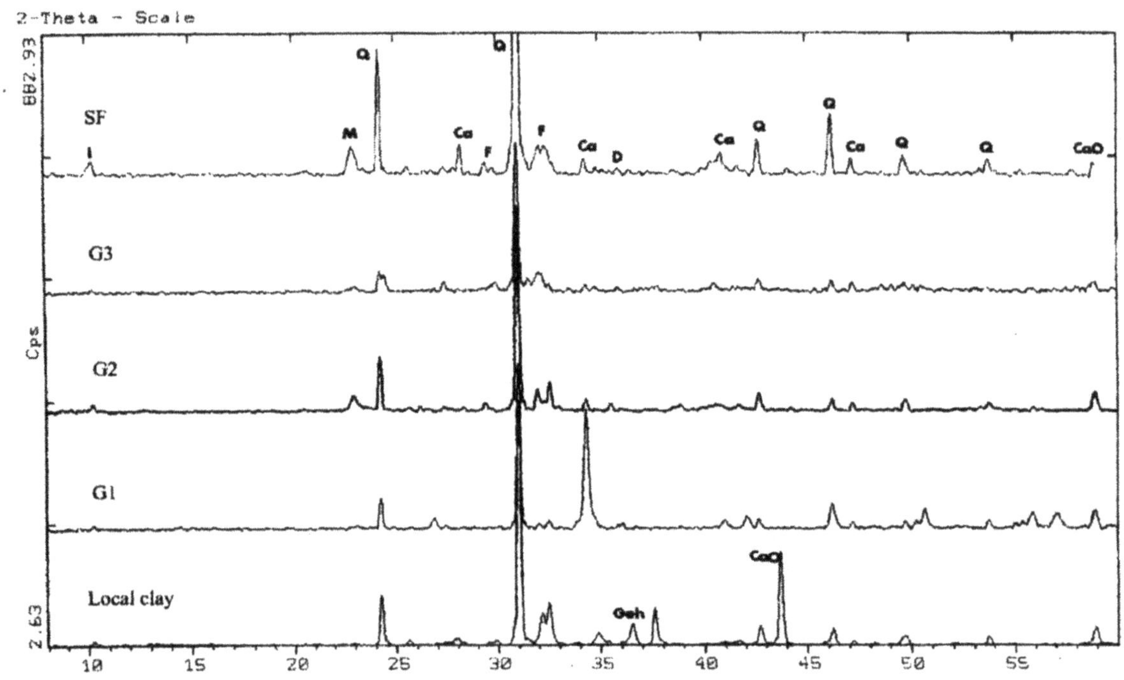

Figure 5. Comparative X-ray diffractions of local clay samples fired at different temperatures. At 400°C, Kaolinite (K) has almost entirely disappeared while a new mineral Gehlenite (Geh) has appeared at 850°C. Ca= calcite, CaO= Calcium oxide, D= Dolomite, Geh= Gehlenite, F= Feldspar, I= Illite, K= Kaolinite, M= Muscovite, Q= Quartz.

BUILDING PROCEDURES

Experimental studies of the natural clays from the clay outcrops near the sites have demonstrated that when making pots the use of temper is not necessary for the construction of any types of containers although it can facilitate the building procedures (Cassano et al. 1993, 1995). In other words, if clays were adequate enough to be used in their natural state, it was easier to build a vessel with intentionally tempered clay, especially in the case of larger pots with thicker walls, which tend to collapse if not supported by an internal skeleton provided by temper. Since the coarser pastes (especially G1 and G2) correspond to the larger vases, one may conclude that Neolithic potters were aiming at minimizing the building risks.

Preparation of the coarse wares G1 and G2 included the addition of inclusions, although in a rather arbitrary fashion. Both types are made with the same clay body, containing the same approximate quantity, size and shape of temper. However, they do exhibit an important intra-type variation, which presents no direct correlation with the vessel form or volume (Plate 2a, b). Larger vases may contain less inclusions and/or smaller grains than smaller pots while the opposite is equally true.

On the other hand, fine ware SF contains finer inclusions than the other groups (Plate 2c). Potters appear to have selected a finer portion of sand to use as temper. In some cases, however, the clay body used to build these fine vases is very similar to the one used for the construction of coarse wares G1 and G2. It appears then that although preparation of the clay bodies of these three groups (G1, G2 and SF) did follow certain rules, they were not very strict. This suggests that potters were satisfied with a rough approximation within a set of accepted rules; a behavior often observed in many contexts of domestic ceramic production (Rye and Evans 1976; Deboer and Lathrap 1979; Rye 1981; Arnold 1985; Miller 1985; Gosselain 1995). Such a situation is most likely a consequence of personal choices beyond technical and/or functional constraints. After all, not all potters had the same skills and/or the same potting experience and therefore had to adjust the preparation of their pastes to their level of competence.

Following socially accepted or traditional rules must have also been an important factor in the decision making process. Ethnographic data from Africa show that more than two thirds of the potters add temper to their clays with no correlation to the subsequent building procedures or intended use of the pots (Gosselain 1995). In some cases, the initial removal of inclusions is even followed by the addition of sand (David 1983)! This situation however is also known for Gaulo-roman domestic ceramics where non-plastics were initially removed from naturally tempered clays and then differentially added to the pastes depending on the potter's needs (Velde and Druc 1999: 80-82).

The case of the third coarse ware G3 is different than the other two. If local sand also constitutes most of the added temper, potters have also systematically added coarser chert inclusions with a mostly subangular to angular shape (Plate 2d). These flakes were probably the by-products of the transformation of the local chert nodules, which were used in the production of most lithic tools on the sites (Giampietri and Tozzi 1990). Such a specific and expedient strategy is also known from the Greek Neolithic site of Franchthi where andesite flakes were added to a specific type of ceramics. Andesite, a non-local rock that came from several hundred kilometers away, was intentionally brought to the site to make stone tools (Vitelli 1984).

Vases of this G3 group are usually small to medium sized with thick walls. They often have a heavy base attached to the vase. None of these characteristics requires the use of temper. On the contrary, small pots with thick walls are the easiest ones to build without adding inclusions. In this case, the building procedures and the weight of tradition can only partially explain such behavior and we must seek other explanations.

FUNCTIONAL ASPECTS

If technical, social and personal choices were responsible for part of the technological variation observed in each category of vases, we must also look at their intended use in order to assess their functional constraints or requirements. In other words, while enjoying certain flexibility in their preparation and building options, potters were also looking at making vases that could fulfil specific purposes.

The mineral composition of the coarse wares would make all of them good cooking pots since the presence of numerous coarse inclusions provides good thermal shock resistance. Quantity and size of inclusions are directly responsible for the increased porosity of clay, which reduces the internal stresses of pots when subjected to rapid changes in temperature (Echallier 1984: 17).

On the other hand, finer paste like that of the SF group, provides a high resistance to breakage from mechanical stress the kind of stress that might result from a vessel being accidentally dropped or knocked over. Fine ware containers would be more suited for non-cooking tasks. With their generally open shape and overall sophisticated decoration - especially on the inside walls- they were more likely to be used either in ritual contexts or for serving purposes.

The nature of inclusions also plays an important role in heat conductivity. Laboratory studies have demonstrated that inclusions with a low thermal expansion coefficient or with one similar or close to that of low fired pottery such as calcite, feldspars or crushed pottery, offer a better resistance to thermal shocks. In contrast, inclusions like quartz which have a high expansion volume when heated do not resist thermal shocks as well and would thus constitute a major drawback for cooking activities (Grimshaw 1971: 791-805; Shepard 1980: 126; Steponaitis 1983: 37-39; Rice 1987: 231).

Calcite and feldspar inclusions have been added to all wares and thus must have improved their thermal conductivity; quartz however, which is also contained in the sand, was added as well in its natural form. We therefore face an apparently contradictory situation on the technological level, but a totally coherent one on the practical level. In this respect, it is interesting to note that the same situation is also found for prehistoric ceramics from Great Britain. For several millennia, sand (which always contains large quantities of quartz) was systematically added to clays that were used to build cooking pots (Woods 1986: 163).

Considering the size and quantity of temper, containers of the G3 group would have been the most adequate cooking vessels since they contain the largest and most numerous inclusions. Moreover their small to medium size make them easy to handle over the fire while their straight sided walls or closed shape would prevent liquids from overflowing and reduce evaporation. Numerous firing and soot traces tend to confirm this function.

As far as their mineralogical composition is concerned, G1 and G2 containers would have also been adequate for cooking since they are both characterized by a good thermal shock resistance and heat conductivity. The great diversity of sizes and shapes of the G1 containers would suit cooking and processing activities, and many of them show firing traces, which tend to confirm this use.

G2 vessels however are usually large or very large which make them very difficult to handle and not suited for cooking. Moreover, they show no traces of firing and have a very high permeability rate. Therefore, despite their good thermal properties, they most likely served as storage containers. In this case there is no direct relation between the physical properties of the ware and the vessel use.

If the mineral composition of the G1 and G3 wares is only partially suited for their function as cooking pots, some of their morphological characteristics do not correspond to the accepted criteria of efficiency either. Vessels intended for cooking are expected to be relatively thin walled to conduct heat better and reduce the thermal gradient between the surfaces; further, they are likely to have rounded contours to permit greater exposure to the heat (Rice 1987: 237). G1 containers however have thick walls and flat bases while most G3 vessels exhibit a heavy foot-like base and even thicker walls. We are thus facing once more the same contradiction between the notion of efficiency and technical choices made by prehistoric potters. This situation, however, has already been noticed for other prehistoric assemblages. Cooking vases from Neolithic sites in Great Britain and in the Near-East (Iran) were also characterized by flat bases and thick walls (Henrickson and McDonald 1983: 637; Woods 1986: 159).

CONCLUSIONS

In recent years, considerable efforts have been devoted to characterize the technological features of pottery and many archaeometric analyses have focused their interests on the functional aspects of vases. The results obtained from petrographical and X-ray diffraction analyses performed on the ceramic assemblages of two Early Neolithic sites from South-eastern Italy, however, cast doubts on the efficiency model according to which the primary consideration for pottery making is to produce suitable physical properties in order to fulfil as efficiently as possible a specific function.

Local sand was added to local clays in the preparation of the ceramic pastes. The mineral composition of this tempering material however, is the same as the one naturally present in the clays and only differs in quantity, size, and in the case of quartz, in angularity. Although these characteristics tend to facilitate the construction of the larger vases, they do not offer a major advantage for the building procedures since all types of pots can be made using the clays in their natural state. On the other hand, pastes used to make finer and coarser vessels can be very similar suggesting that, within a set of accepted rules, potters were not very strict in the preparation of their pastes and adapted them to their own needs. On the other hand inclusions of feldspars and calcite present in the added sand could offer a better resistance to thermal shocks and thus be functionally appropriate for cooking vases. Quartz, however, which is a major component of the added sand, was also systematically added to the pastes despite the fact that it represents a technological drawback because of its low resistance to thermal shocks.

This apparently contradictory situation suggests that theoretically incompatible behavior can be practically coherent or at least may be viewed as such by a group of people and thus accepted and practiced as the right way for pottery making. On the other hand, within a socially accepted set of technological rules, individual choices can also be an important factor responsible for the observed variation. Thus we see that technical choices can be explained equally in terms of culturally learned behavior and functional strategies.

Technological studies then should be aimed at exploring the interaction between materials, tools and gestures in order to deal with all the decision making processes that underlie ceramic production. In this sense, ceramic technology is more than the sum of the necessary strategies used in the construction of an object, but it is also the result of human choices and the implementation of a specific know-how.

Acknowledgments
The writer is indebted to Giovanni Boschian from the Dipartimento di Scienze Archeologiche, Univesità di Pisa, for his valuable collaboration with the petrographical analysis. Many thanks also to Michel Preda from the Département des Sciences de la Terre de l'Université du Québec à Montréal for performing the X-ray diffraction analyses. This research has been made possible with the financial support of the Social Sciences and Humanities Research Council of Canada. Finally the author wishes to thank Claude Chapdelaine and Adrian Burke from the Département d'Anthropologie, Université de Montréal, for useful comments on earlier drafts of this article; however, the writer alone takes full responsibility for the ideas expressed herein.

References

Arnold, D. E. 1985. *Ceramic theory and cultural process.* Cambridge: Cambridge University Press.

Barker, G. 1981. *Landscape and society. Prehistoric Central Italy.* New York: Academic Press.

Barker, G. 1985. *Prehistoric farming in Europe.* Cambridge: Cambridge University Press.

Balenzano, F., Dell'Anna, L., and Di Piero, M. 1977. Richerche mineralogiche, chimiche e granulometriche su argille subappennine della Daunia (Puglia). *Geologia Applicata e Idrogeologia.* 12: 33-55.

Balenzano, F., Dell'Anna, L., and Di Piero, M. 1984. Argille pugliesi: raccolta dei dati sulla composizione mineralogica. In *Indagini sui materiali di cava (calcari, calcarenti e argille) della Puglia ai fini di favorirne l'utilizzazione. La valorizzazione e la possibilita di nuovi procedimenti di lavorazione ed estrazione,* ed. GEO. s.r.l., Assessorato all'industria, al commercio e all'artigianato, Bari, pp. 33-83.

Bradford, J. S. P. 1949. "Buried landscapes" in Southern Italy. *Antiquity* 21: 58-72.

Bradford, J. S. P., and Williams-Hunt, P.R. 1946. Siticulosa Apulia. *Antiquity* 20: 191-200.

Braun, D. P. 1983. Pots as tools. In *Archaeological hammers and theories,* eds. J. A. Moore and A S. Keene. New-York: Academic Press, pp. 108-127.

Bronitsky, G. 1986. The use of material-science techniques in the study of pottery construction and use. In *Advances in archaeological method and theory* 9: 209-276.

Cassano, S., Eygun, G., Garidel, Y., and Muntoni, I. 1993. La produzione ceramica nel Neolitico del Tavoliere. Spunti da uno studio sperimentale. *Atti del 15 Convegno Nazionale sulla Preistoria, Protostoria e Storia della Daunia.* San Severo: Gerni., pp.41-56.

Cassano, S., Eygun, G., Garidel, Y., and Muntoni, I. 1995. Pottery making in Southern Italy Neolithic: An experimental study. In *Estudis sobre ceramica antiga. Proccedings of the European Meeting on Ancient Ceramics,* eds M. Vendrell-Saz, T. Pradell, J.Molera , and M. Garcia. Barcelona: Generalitat di Catalunyat, pp. 11-16.

Chapman, J., and Müller, J. 1990. Early farmers in the Mediterranean basin: the dalmatian evidence. *Antiquity* 64: 127-134.

Ciancio, A., Dell'anna, A., and Laviano, R. 1994. Indagine chimico-mineralogiche su ceramica a pasta grigia proveniente dalla Puglia Centrale. In *Ceramica romana e archeometria: lo stato degli studi,* ed. G. Olcese. Firenze: Edizioni all'insegna del Giglio, pp. 261-276.

David, N. 1983. Poterie domestique et rituelle du Sud-Bénin: etude ethnoarchéologique. *Archives Suisses d'Anthropologie Générale.* 47(2): 121-184.

Deboer, W. R., and Lathrap, D.W. 1979. The making of Shipibo-Conibo ceramics, In *Ethnoarchaeology: implications of ethnography for archaeology,* ed. C. Kramer. New-York: Columbia University Press, pp. 102-138.

Dell'Anna, A. 1986. Applicazione dell'analisi mineralogica allo studio della ceramica impressa di c.no San Matteo-Chiantinelle (FG). *Taras* 6: 97-103.

Dell'Anna L., Di Pierro, M., Nuovo, G., Ciaranfi, N., and Ricchetti, G. 1974. Giacimenti di argille ceramiche - Puglie. In *Giacimenti di argille ceramiche in Itali,* eds. F. Veniale and C. Palmonari, Gruppo Ital.iano de l'Association internationale pour l'étude des argiles, pp. 195-223.

Dondi, M., Fabbri, B., and Laviano, R. 1992. Characteristics of the clays utilized in the brick industry in Apulia and Basilicata (Southern Italy). *Mineralogia. Petrografia. Acta* 35: 181-191.

Echallier, J.C. 1984: *Éléments de technologie céramique et d'analyses des terres cuites archéologiques.* Salon, France: Documents d'archéologie méridionale, Méthodes et Techniques 3

Eygun, G. 1998. *Les productions céramiques du Néolithique ancien du Tavoliere (Ripa Tetta et Lagnano da Piede, Italie du Sud-Est: Technologie et implications socio-culturelles.* Dissertation thesis. Paris: Laboratoire de Préhistoire et technologie, Université de Paris-X.

Eygun, G. 1999. Des vases et des gens: interprétations technologiques d'analyses céramiques. *Paléoquébec* 29: 45-52.

Giampietri, A., and Tozzi, C. 1990. L'industria litica del villagio di Ripa Tetta (Lucera). *Atti del 11 Convegno Nazionale sulla Preistoria-Protostoria della Daunia,* San Severo, December 1989, pp. 72-86.

Gosselain, O. 1995. *Identités techniques: le travail de la poterie au Cameroun méridional.* Dissertation thesis. Bruxelles: Faculté de Philosophie et de Lettres, Université Libre de Bruxelles.

Grim, R. E. 1968. *Clay mineralogy.* New York: McGraw Hill.

Grimshaw, R. W. 1971: *The Chemistry and Physics of clays.* Vermont: Techbooks.

Guilaine, J. 1980. Problèmes actuels de la néolithisation et du Néolithique ancien en Méditerranée occidentale. In *Interactions and acculturation in the Mediterranean,* eds. J. G. P. Best and N.M. De Vries. Amsterdam: B.R. Grüner, pp. 3-22.

Henrikson, E. F, and. MacDonald, M.A. 1983. Ceramic form and function: an ethnographic search and archaeological application. *American Anthropologist* 85: 630-643.

O'Brien, M.J., Holland, T.D., Hoard, R.J. and Fox, G.L. 1994. Evolutionary implications of design and performance characteristics of prehistoric pottery. *Journal of Archaeological Method and Theory* 1(3): 259-304.

Miller, D. 1985. *Artefacts as categories: a study of ceramic variability in Central India.* Cambridge: Cambridge University Press.

Muntoni, I. 1995. A technological approach to the Neolithic pottery production at Masseria Candelaro (FG - Italy). In *The Ceramics Cultural Heritage,* ed. P. Vincenzini. Techna - Monographs in Materials and Society, no.2, pp. 301-308.

Rice, P. 1987. *Pottery Analysis: A Sourcebook.* Chicago: The University of Chicago Press.

Rye, O. S. 1981. *Pottery Technology: Principles and Reconstruction*. Washington D.C.:Taraxacum.

Rye, O. S., and Evans, C. 1976. *Traditional pottery techniques of Pakistan: field and laboratory studies*. Smithsonian Contributions to Anthropology, No.21. Washington, D.C.: Smithsonian Institution Press.

Schiffer, M. B. and Skibo, J. M. 1987. Theory and experiment in the study of technological change. *Current Anthropology* 28: 595-622.

Searle, A.B., and Grimshaw, R.W. 1959. *The Chemistry and Physics of Clays and Other Ceramic materials*. New-York: Interscience.

Shepard, A. O. 1980. *Ceramics for the Archaeologist*. 5th edition. Washington D.C.: Carnegie Publication 609.

Steponaitis, V. P. 1983. *Ceramics, Chronology, and Community Patterns: An Archaeological Study at Moundville*. New York: Academic Press.

Van der Leeuw, S. 1993. Giving the potter a choice. Conceptual aspects of pottery techniques. In *Technological choices. Transformation in material cultures since the Neolithic*, ed. P. Lemonnier. London: Routledge, pp. 238-288.

Velde, B., and Druc, I. 1999. *Archaeological ceramic materials*. Berlin: Springer Verlag.

Whitehouse, R.1991. Social organisation in the Neolithic of southeast Italy. In *Early settlements in the Western Mediterranean islands and their peripheral areas*, eds W. H. Waldren, R. Chapman, J. Lewthwaite, and R. C. Kennard. Oxford: British Archaeological Reports, international Series 229, pp. 109-133.

Whitehouse, R. 1992. Underground religion: cult and culture in prehistoric Italy. *Accordia Specialist Studies on Italy, vol. 1*, eds. E. Herring, R. D. Whitehouse, and J. B. Wilkins. London: Accordia Research Centre, University of London, pp. 216.

Vitelli, K. 1984. *Social implications of the EN pottery assemblage at Franchthi cave*. Paper presented at the Society for American Archaeology, Los Angeles, April 17 1984.

Woods, A. J. 1986. Form, fabric, and function: some observations on the cooking pot in Antiquity. In *Ceramics and Civilization: Technology and Style*, ed. W. D. Kingery. Vol. II. The American Ceramic Society, Colombus, pp.147-172.

PLATES

PLATE 1 – CHAPTER 4

Sample ET4: compositional maps globule (X-ray map for K alpha emission), in false colors for Na, Mg, Al, K, Ca, Fe, Si and C around a relatively big carbon rich inclusion. The lowest right figure corresponds to a BSE image. In the element maps lighter is the area, higher is the content in the element. The lightest zones in the C-map correspond to pure carbon. In the Na-map the light zone corresponds to an albite crystal. The Mg and Fe are concentrated in the same are that may be correlated with the K one showing the occurrence of phyllosilicates. Alumina is present nearly in the whole area and as a corona around the carbon drop as well as silica, and a few Ca, K, Mg, Na. Analysis are shown Table 4. The acquisition time is one hour.

PLATE 2 a, b, c, d – CHAPTER 5

Plate 2a. Thin section of Grossolana 1/2.(G1/G2) paste. It contains numerous quartz and calcite inclusions. XPL (X 50).

Plate 2b. Thin section of Grossolana 1/2 (G1/G2) paste. It contains coarser calcite inclusions than in the previous group. XPL (X 50).

Plate 2c. Thin section of Semi-fine (SF) paste. It contains a finer fraction of the same inclusions as those contained in G1/G2 paste. XPL (X 50).

Plate 2d. Thin section of Grossolana 3 (G3) paste. It contains coarser and more angular inclusions of quarzt and chert. XPL (X 50).

PLATE 2 e, f – CHAPTER 8

Plate 2e. Microphotography of a ceramic paste (cooking pot) tempered with slate fragments. Huari group 1. LN 25X.

Plate 2f. Microphotography of ceramic paste (cooking pot) tempered with slate fragments. Huari group 2. LN 80X.

PLATE 3 – CHAPTER 7

a: Buff ware bowl. (From the Norton Allen collection, now housed at the Arizona State Museum. Photograph by Owen Lindauer.)

b: Typical paste in buff ware thin section (8.25x). Note the sparse schist, sand, clay lumps, and caliche.

c: Sand from Petrofacies E, sample GB-083 (8.25X). Note the different textures and types of schist.

d: Briquette with crushed schist (13X). Note the crushing features on grains, finely disseminated mica, quartz released from the schist, and volcanic grains released from the crushing implements.

e: Grain size distribution: Unimodal from silt to sand (16.5X).

f: Grain size distribution: Bimodal from silt to sand (16.5X).

g: Clay lumps in buff ware paste (16.5X). Note the homogeneity of the clay lumps compared to the heterogeneity of the surrounding paste.

h: Early pottery made with Santa Cruz River clay (16.5X). Note the heterogeneity of the paste, including the clay lumps.

PLATE 4 – CHAPTER 7

a: Buff ware in which silt and sand have the same composition (16.5X). Note the volcanic grains, quartz, and plagioclase in the sand size fraction.

b: Buff ware in which silt and sand have the same composition (66X). Note the volcanic grains, quartz, and plagioclase in the silt size fraction. This image is focused on the center of Plate 4.A but is rotated 90 degrees).

c: Buff ware in which silt and sand have different compositions (16.5X). Note the volcanic grains, caliche, and granitic rock fragments.

d: Buff ware in which silt and sand have different compositions (66X). Note that quartz and potassium feldspar from granitic rocks dominate the silt fraction.

e: Pasil volcanic clay pottery (16.5X). Note that plagioclase and hornblende grains dominate the sand size fraction.

f: Pasil volcanic clay pottery (66X). Note that plagioclase and hornblende grains are clearly present in the silt size fraction.

g: Tucson basin volcanic-sand-tempered pottery (16.5X). Note the volcanic rock fragments, plagioclase, and quartz in the sand size fraction.

h: Tucson basin volcanic-sand-tempered pottery (66X). Note the clear presence of quartz, plagioclase, and volcanic rock fragments in the silt size fraction. This image is focused on the plagioclase grains at the center top of Plate 4.G, but is rotated 90 degrees.

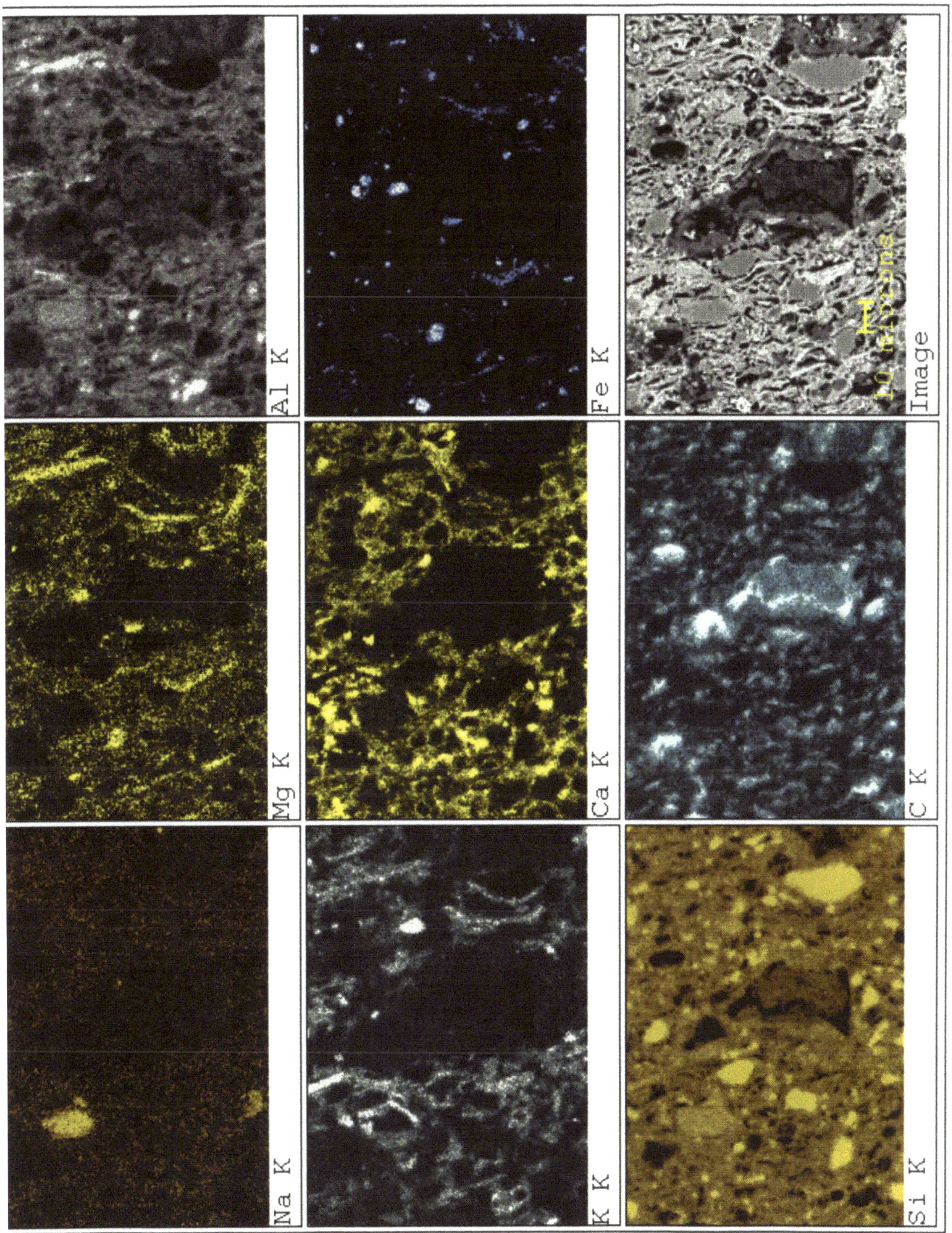
Al K
Fe K
Image
10 microns
Mg K
Ca K
C K
Na K
K K
Si K

PLATE 2

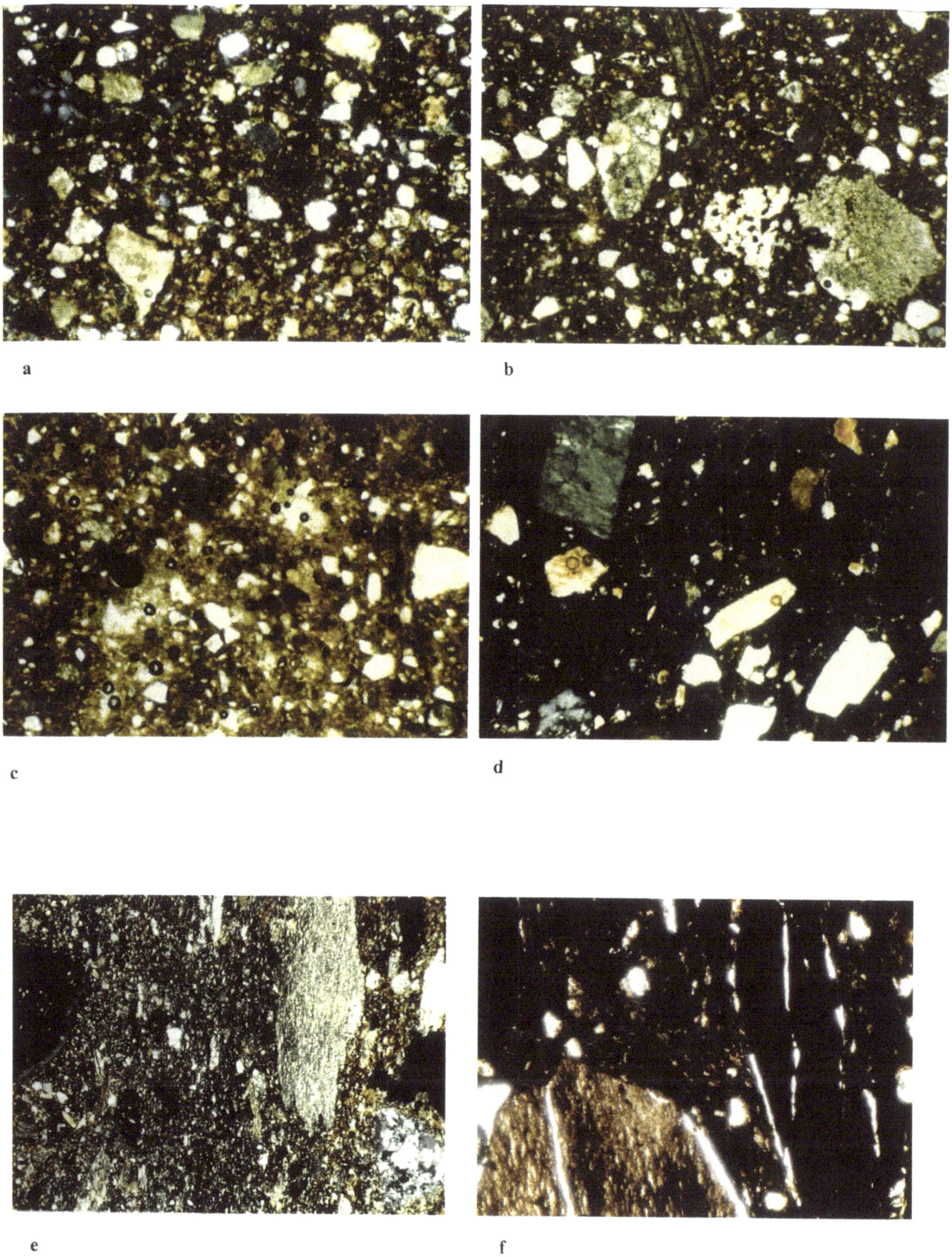

a b c d e f

PLATE 3

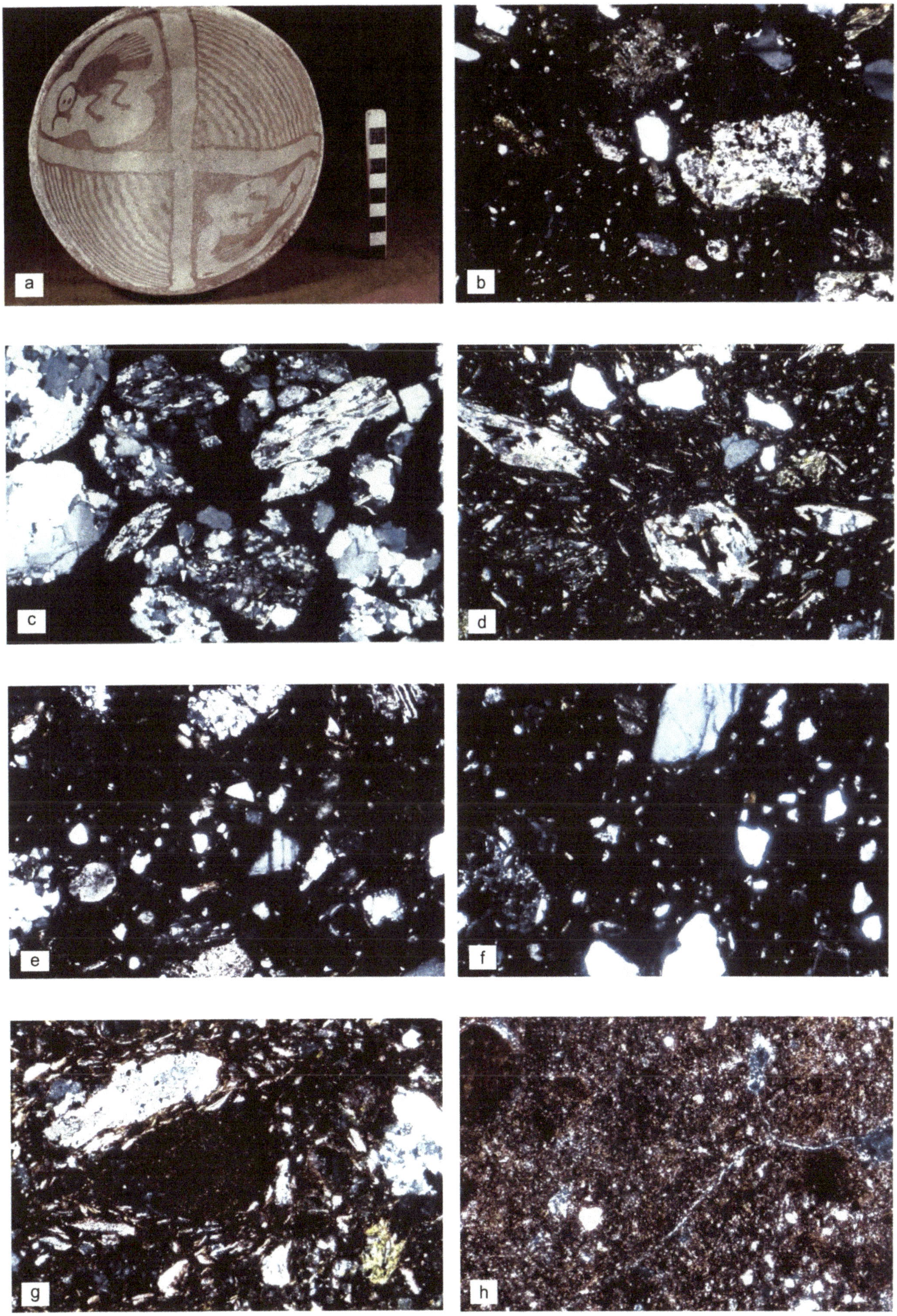

PLATE 4

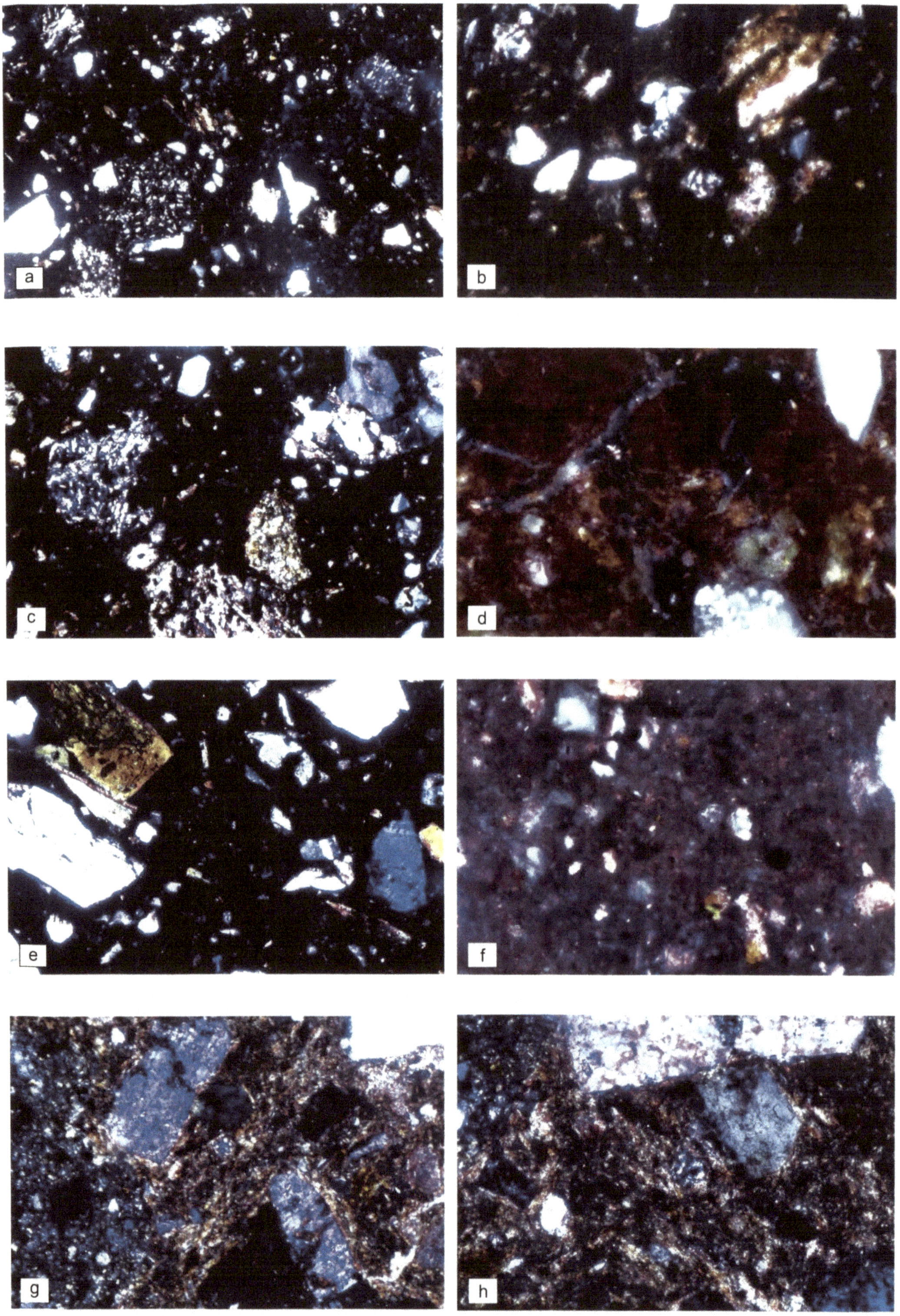

CHAPTER 6

GRAIN DISTRIBUTION BY IMAGE ANALYSIS OF THIN SECTIONS IN SOME GAULO-ROMAN COMMON WARE, ST MARCEL (INDRE) FRANCE

Isabelle Bouchain
UMR Archéologie et Sciences de l'Antiquité, Université de Nanterre, Paris

Bruce Velde
Laboratoire de Géologie École Normale Supérieure, Paris
Corresponding author: papa@geologie.ens.fr

Thirty one samples of common ware ceramics coming from the Gaulo-Roman site at Saint Marcel (Indre) France were observed using image analysis of ceramic thin sections. The objects were found in contexts ranging from 1st to 5th centuries. Several patterns of grain size distribution suggest the different uses of clay and grit resources to produce the ceramics. Use of natural river bank sediments gives smooth log-normal distribution of grain sizes. Some of the clay resources were tempered by larger grits giving high and irregular abundance of these grains in the coarse fraction. Use of a sieved grit source with river sediment gives high abundance of larger grits with a sharp maximum value. It appears that different methods of producing paste for common ware ceramics were used over the period investigated. Correlation of paste type and ceramic objects suggests that a variety of pots from cooking ware to storage to table ware were produced by each of the different methods. Hence apparently different producers selected different methods of producing a range of similar products.

INTRODUCTION

Optical characterization of grits in ceramic bodies can be used to identify the types of materials used to form the pots produced (Woods 1985; Mason 1995; Whitbread 1991, Betts 1991). It is also possible to use such information to determine the methods of producing the paste. Hence, a careful examination of the clay - coarse grain relations can give information not only concerning the source materials but also the techniques used by potters.

An estimation of the relative grain size distribution of grits in ceramic sherds can be made by observing thin sections using current image analysis methods (Maggetti et al. 1988). In the use of such methods it is evident that a two dimension sampling (thin section image) of a three dimensional array of grains (sherd) will give random sections of different grains, only a few of which will show the maximum and exact diameter of a grain (see Velde and Druc 1999). Nevertheless, if the same method is used to estimate relative population characteristics, one can compare samples with one another in the same population without worrying about the absolute values of grain size.

The potential of grain size estimation for the characterization of grits in a ceramic is great, but up to now few studies have been designed to compare different ceramic types in order to describe methods of production or differences in origin of the components in a given archaeological context of ceramic production. Some studies have shown the potential of such methods (Maggetti et al. 1988; Whitbread 1991; Fieller and Nicholson 1991; Matthew et al. 1991, for example). Also one can obtain grain size distribution by point counting methods as advocated by Miksa (1998).

We propose to explore the consequences of some observations of grain size distribution using image analysis of photographs of thin section of sherds which were treated by a computer based method. The different ceramic objects observed in this study are very homogeneous, being common ware objects of similar types (cooking pots, storage containers and plates and other table ware) produced over a 500 year period at the same site with apparently the same clay sources

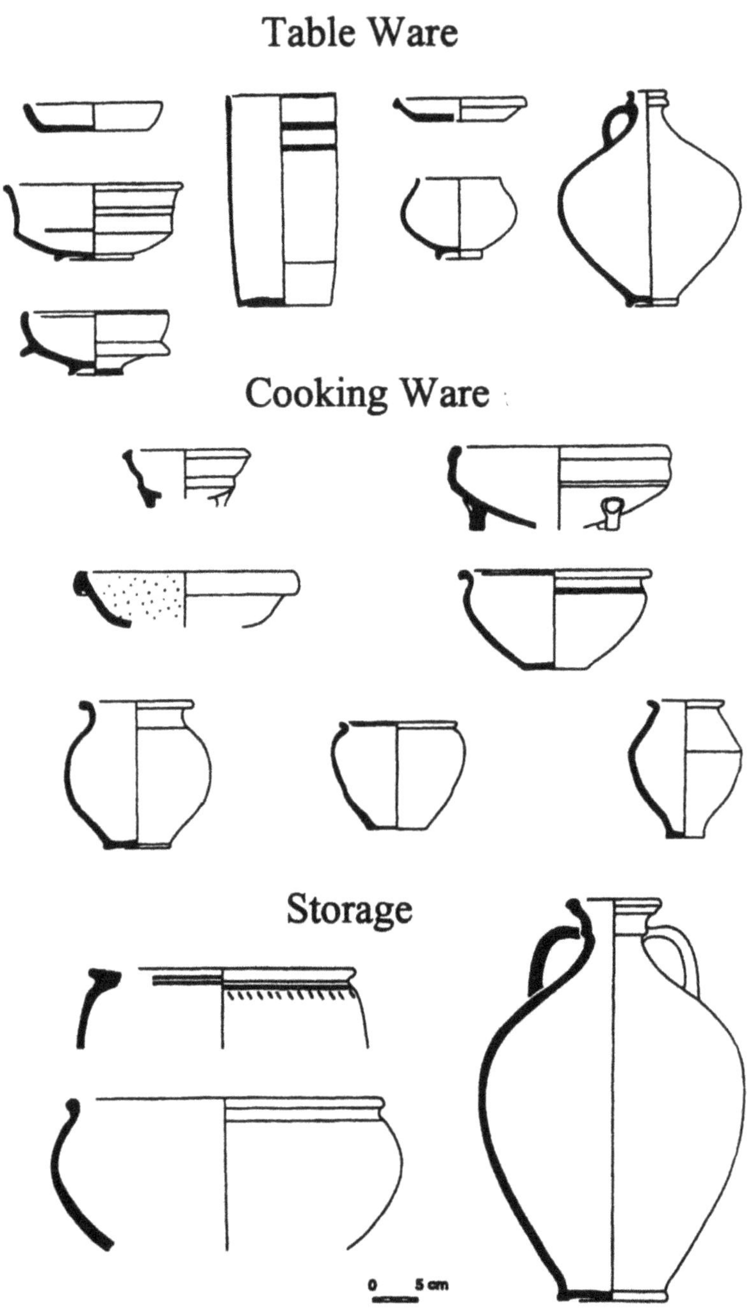

Figure 1. Representation of common ware types encountered at Saint Marcel.

METHODS AND MATERIALS

Samples and Site

The sherds investigated here come from a programmed archaeological investigation which has been active for 10 years (see Dumasy et al. 1997). They represent common ware ceramic vessels examples of which are shown in Figure 1.The Gaulo Roman town of Argentomagus is next to the village of Saint Marcel which is on a plateau 2 kilometers from the town of Argenton sur Creuse which is situated on either side of the Creuse River. Both are on sedimentary Cretaceous limestone which also forms the basement of the river bed. All of the soils and sediments are of limestone-derived materials except a narrow strip (tens of meters) of micaceous sandy material found on either side of the Creuse River.

The common ware samples studied here have been initially described by Bouchain (1992). Thin sections were made of 100 samples covering a range of the materials catalogued from the archaeological discoveries. All of these samples showed a very similar identifiable mineralogical assemblage which was comprised of quartz, potassium feldspar, white mica and biotite with occasional rare accessories of iron oxide concentrations, and very rare metamorphic pyroxene, kyanite, and garnet minerals. This mineralogy suggests a two mica granite as a pre-cursor. The almost identical grit mineralogy suggesting a common origin of the clay and grit resources used to form the pots. These resources seem to have been derived from the sandy micaceous deposits along the bed of the Creuse river just one kilometer from the site at St Marcel. The potential origin of the minerals found in these sediments is up river (Figure 2) where the Creuse cuts a crystalline two mica granite massif 40 km from the site.

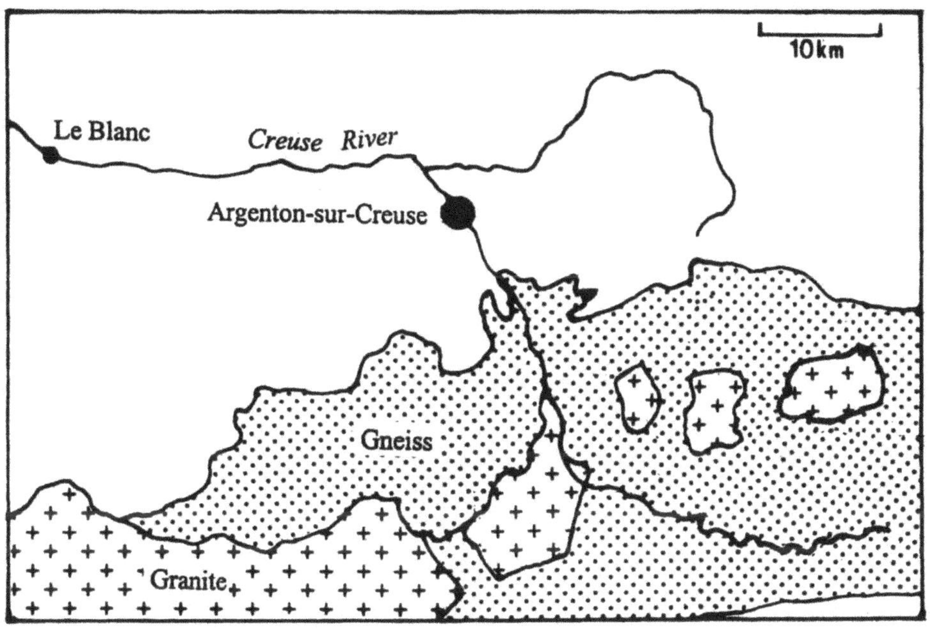

Figure 2. Location map and abbreviated geology of the area around Argenton sur Creuse (Saint Marcel). The non-shaded area is sedimentary, largely limestone basement.

OPTICAL OBSERVATIONS

Thirty two thin sections made from the sherds (30 µm thickness) were photographed using a normal petrographic microscope where the photograph covered a 2 by 3 mm sector of the sherd. The contrast between the brown clay sized fraction and the translucent grits (essentially quartz and feldspar grains plus the white mica muscovite) was such that these mineral grains could be isolated by thresholding of scanned grey level images of microphotographs using normal image analysis programs. Repeated thresholding of the same image showed that differences in total area selected as being grits varied by less than 10 percent of the grit area for samples with nominal total area of 20%. We assume that the method is reasonably reproducible and can be used on a comparative basis for materials similar to those investigated.

IMAGE ANALYSIS

The thresholded image (binary) of the 32 thin sections was treated using a program developed by E. Moreau of the University of Poitiers (see Moreau et al. 1999). The estimation of grain sized was made by an opening (erosion-dilation) image analysis procedure which compares the surface area of all grains eliminated by erosion cycles. Thus the importance of grains size is estimated by percent of the area of the photograph occupied by each size category.

Table 1 gives the modes, maximum grain diameters and other grain size distribution data as well as the function of the ceramics (table ware, cooking ware and storage vessels).

Table 1. Image analysis and archaeological data for the samples investigated. Total area indicates the area of the thin section which was identified as grit material. %>120 = the area of grains greater than 120 μm in thin section. 90-120 μm = the area of grains in this size fraction. Mode = the most abundant category in the log normal trend of grain sizes (excluding the erratic large grits when present). Distribution = the different types of grain size distributions where 1 = a near log normal distribution, 2 = a log normal distribution with some large grains present but with a distinct maximum value indicating a sieving process, 3 = erratic large grains in rather high abundance. Use of the common wares is divided into table ware (t) cooking ware (c) and storage (s) function.

Sample	total area (%)	%>120μm	90 – 120 μm	mode (μm)	distribution	use
1	14	1	1	16	1	c
2	28	7	3	14	2	c
9	15	1	2	17	3	c
12	16	3	2	24	3	c
16	6	0	0	8	1	t
20	17	7	2	16	3	s
21	15	3	2	11	3	c
22	32	12	3	18	3	s
26	15	0	2	4	2	t
34	13	0	0	25	1	t
41	28	0	0	17	1	t
47	13	0	0	20	1	c
49	16	0	1	20	1	c
50	18	0	1	28	1	c
56	18	3	2	30	2	c
57	28	4	1	25	3	s
62	25	7	3	31	2	c
63	24	4	2	21	3	c
64	30	9	3	20	2	c
71	26	13	2	20	2	c
76	13	2	2	21	3	t
77	13	2	2	17	2	t
79	11	1	2	21	3	c
80	10	0	0	21	1	t
82	15	4	2	17	3	c
83	12	2	2	16	3	c
85	13	1	3	26	3	c
90	9	1	1	16	3	t
92	6	0	0	19	3	t
93	21	7	2	21	2	s
99	27	4	3	30	3	c
102	36	10	5	30	2	c

EXPERIMENTAL RESULTS

The image analysis of the grains in the sherds can be represented on area percent (analogous to weight percent) - grain size coordinates (Figure 3a). The size categories the most represented are the smaller ones which give a gradually decreasing distribution to larger grain sizes. If one uses the log-linear representation (see Rock 1988; Parkin and Robinson 1992) which indicates a log-normal relation common for distributions of geological materials, one can easily distinguish certain characteristics of the grain size distributions determined from their two dimensional representations (Figure 3b). On the log - linear scale one sees that in many cases the surface area distribution is symmetrical around a mode grain size value. The average value of the grains sizes in these distributions will be greater than the mode, unlike a Gaussian distribution. One quarter of the 32 samples investigated showed this simple distribution indicating a simple sedimentary origin (McLean 1995: 13).

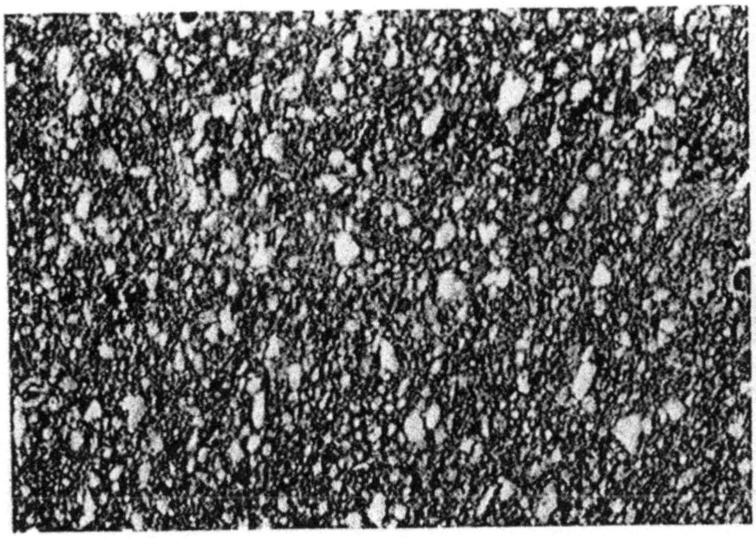

a

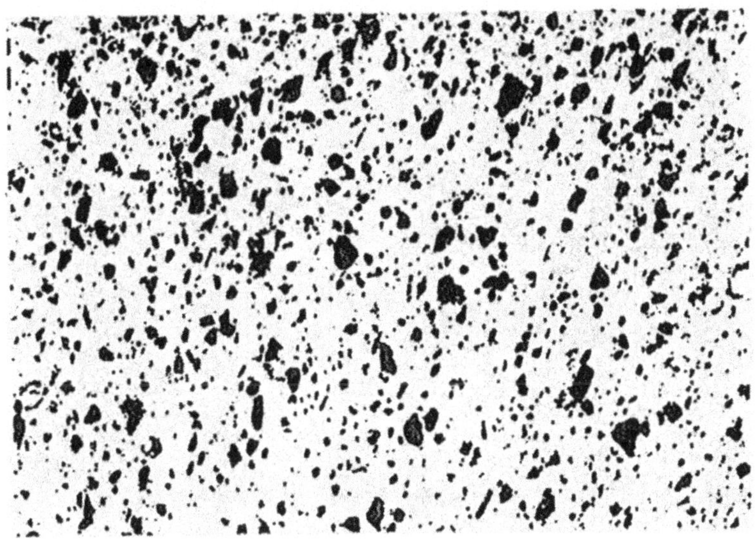

b

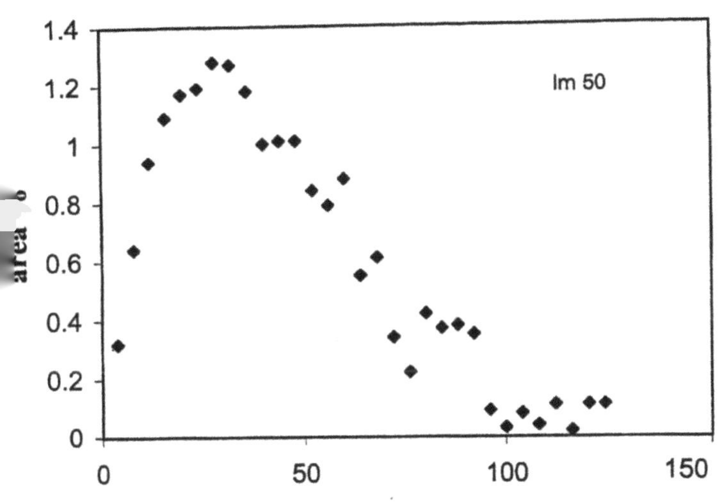

c

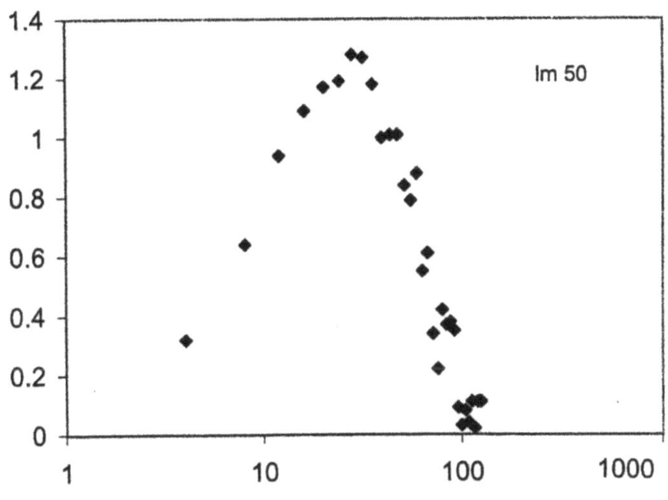

d

Figure 3. a) Grey level photograph of sample Lm 50. b) Binarised (black and white) representation of the grits shown in clear in figure 3a. c) Area percent-grain diameter relations for a sample of a globular storage vessel from Saint Marcel with 18% area occupied by grits in the thin section. d) The same data as in fig 3c, plotted on area percent log grain diameter co-ordinates. This representation gives a log-normal type distribution where the points form a relatively smooth, rather symmetric curve.

In a visual observation of the grain size distribution curves established from the image analysis of the thin section photographs (total area vs. grain diameter) one can identify three major types. **Type 1**: The grain sizes are evenly and symmetrically distributed on log size - area percent plots, (Figure 4). This indicates a single simple origin for the sediments used to make the paste. **Type 2**: A regular distribution is present but one also finds large size **grains**, of irregular distribution pattern that are sharply limited to high values (between 120 and 200 µm) (Figure 5). There is a maximum in the distribution curve (mode) as in case 1, but the abundance of larger grain sizes shows an abrupt maximum cutoff value. **Type 3**: the distribution is regular with a maximum value but the larger grain sizes, which show an erratic, very high representation of some sizes (up to 600 µm in diameter) (Figure 6). In all of the samples investigated, the regular distribution trends have a maximum value near 120 µm with a mode range from 8 to 28 µm.

a

b

c

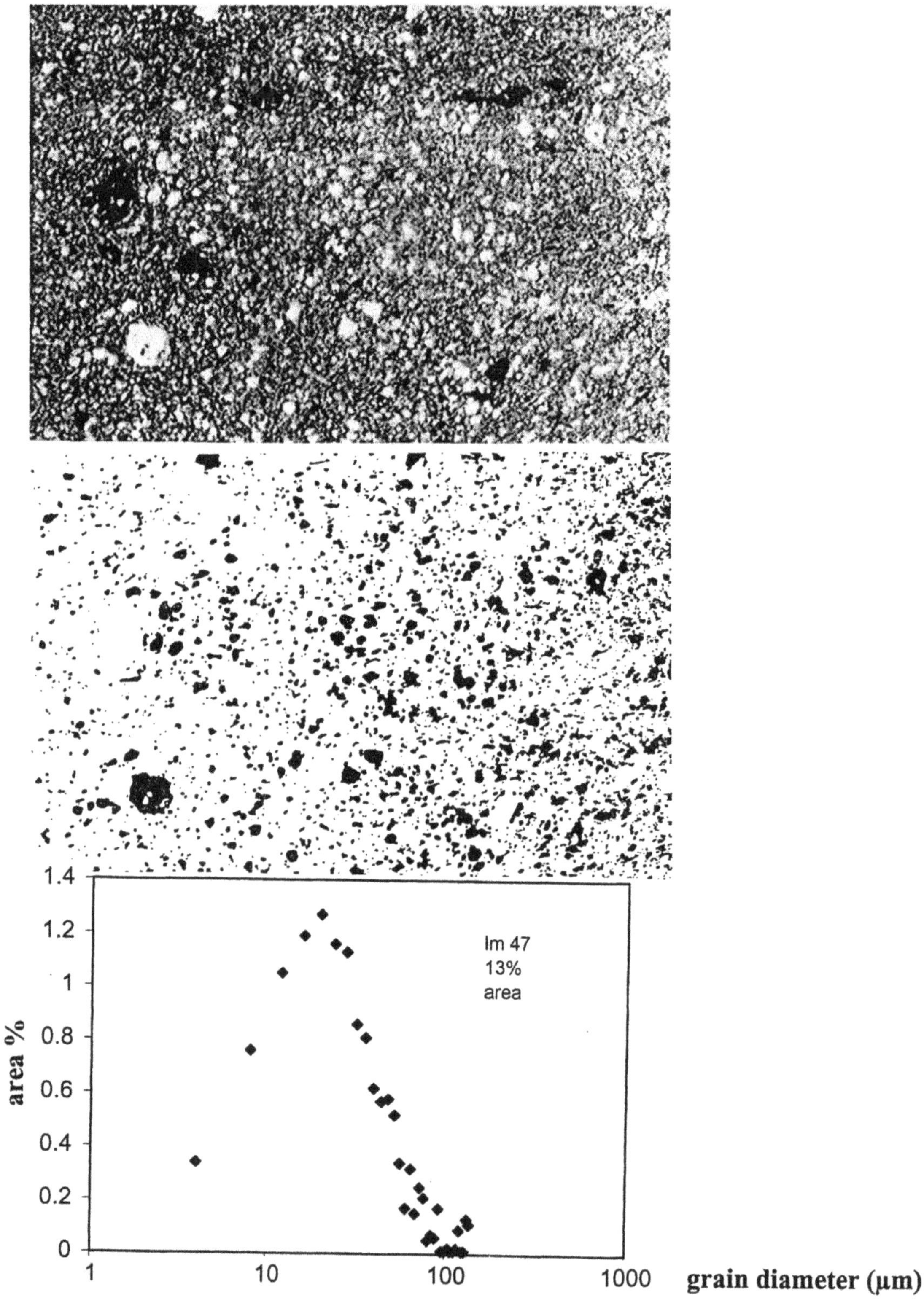

Figure 4. a) Grey level photograph of sample Lm 47. b) Binarised (black and white) representation of the grits shown in clear in fig. 4a. c) Grain size distribution in sample Lm47, a cooking bowl, showing the symmetric, smooth distribution common for natural sedimentary material associated with clays. It has 13% of the thin section area covered by grits. Scale 2 x 3 mm.

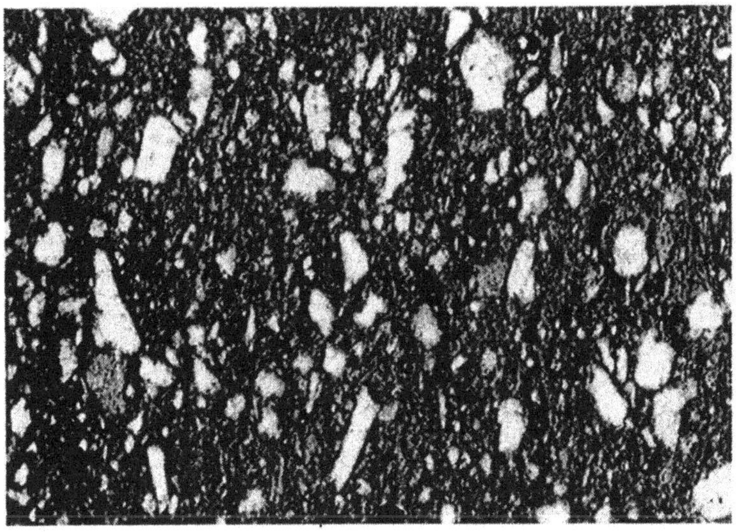

a

b

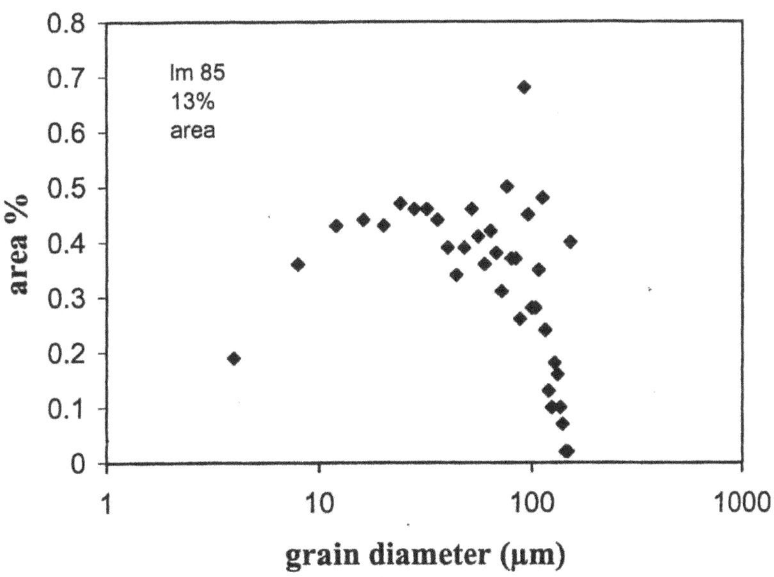

c

Figure 5. a) Grey level photograph of sample Lm 85. b) Binarised (black and white) representation of the grits shown in clear in fig. 5a. c). Grain size distribution for Lm85, from a cooking pot, with the same area of LM47 but a very different distribution pattern. There is a general maximum as in the case of Lm47, but there is a significant number of larger grits present. The larger grits show a distinct cut-off value near 150 um indicating that the larger grits were sized out of the sample, probably by sieving. Scale 2 x 3 mm.

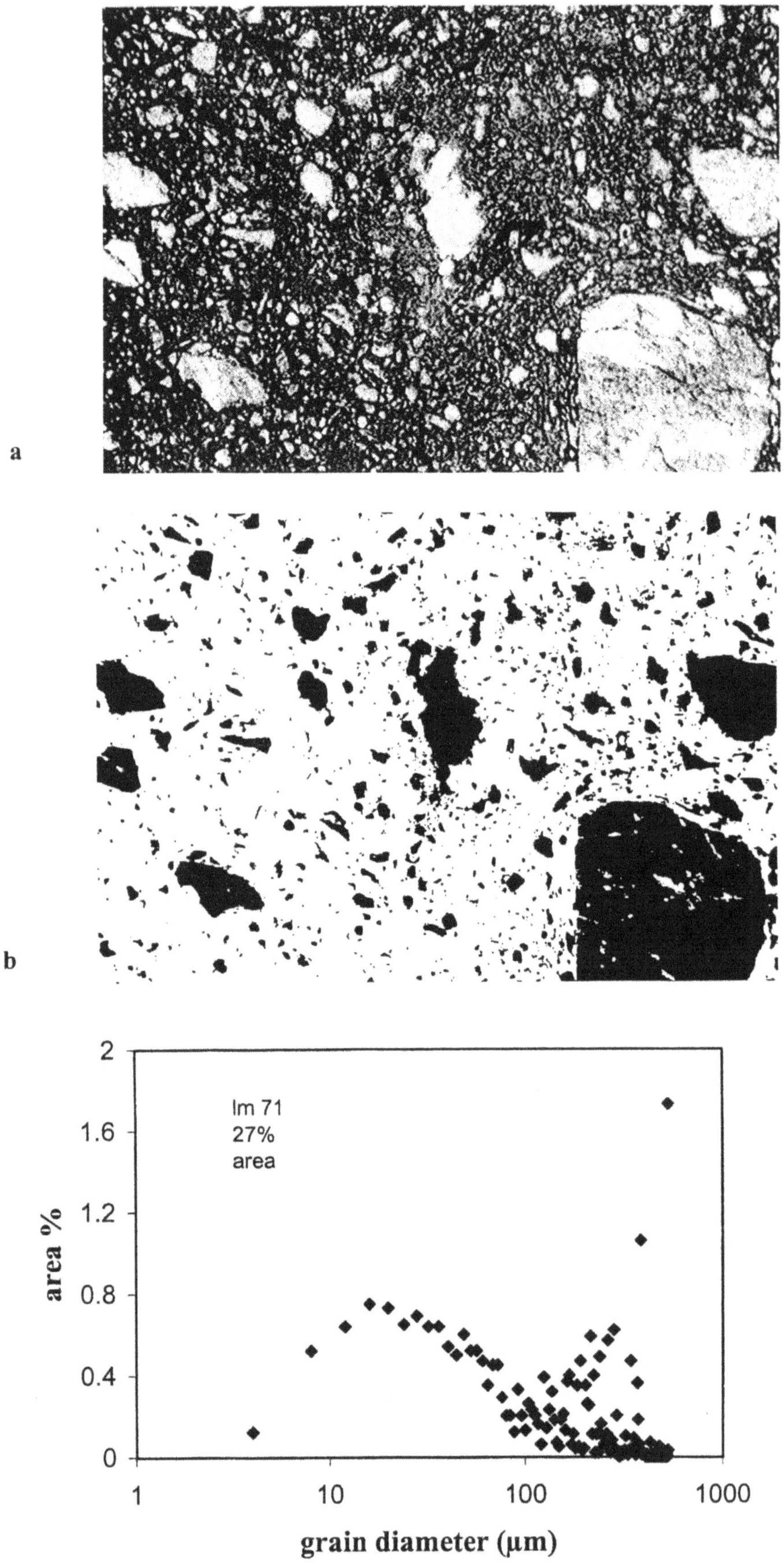

Figure 6. a) Grey level photograph of sample Lm71. b) Binarised (black and white) representation of the grits shown in clear in fig. 6a. c). Grain size distribution for Lm71, a cooking pot, with 27% area of grits and a distinctly greater abundance of very large grits with sizes up to 500 um diameter in the thin section. The large grits show a scatter of very large sizes. Scale 2x3mm.

DISCUSSION

The observations of the grain size distributions established by image analysis indicate several relations. It is clear that the distributions have a similarity in their modes even though the larger grain sizes can show different types of distributions. It appears then that there is an underlying gain size distribution in all samples, similar to that of the simple distributions which would be expected in the use of initially similar sediment sources. These are probably the sediment materials found on the banks of the Creuse River which includes non-clay, micaceous and sand grains. The range in the mode of the simple distribution curves, is from 8 to 28 μm

The presence of larger grains in erratic proportions in some samples indicates an addition of materials foreign to the initial clay-rich resource. These materials could come from different layers of more sand-rich materials in the Creuse River bed. In the samples investigated the modes of these distribution types range only from 16 to 30 μm, very similar to the samples showing a simple distribution curve.

In some samples (type 2 Table 1) there is a clear maximum size for large grains (near 150μm) suggesting that a sizing mechanism (sieving) was used to give a limiting maximum size for these grits. The mode values range from 14 to 31μm., again similar to the simple distribution curve.

The similarity of the mode values for all of the grain size distribution types indicates that a similar source of clay-rich materials was used by the potters of the site for a considerable period of time (about 500 years). If one can identify several types of grain distribution patterns it appears that the same base material was used and treated in different ways. Not only is the grain size the same but also the mineralogy is almost identical coming from a two mica granite materials. The different grain abundance distributions then suggest different uses of the same natural resources combined to make the paste of the pots investigated. Type 1 indicates a use of untreated raw clay-rich sediments used to make the posts. Type 2 indicates the addition of large grains to modify the plasticity of the initial clay resources. The clear maximum grain size here indicates a size selection process. Type 3 indicates use of added grits with an wide range of grain size.

An obvious question which can be posed is whether or not a given ceramic has a different grain size distribution (amount of grits) depending upon its function. The initial thought is that where a cooking function is apparent the object will be subjected to high thermal gradients in a fire which would require specific paste characteristics. Of the five samples which were of tripod or cooking bowl shape the range in mode of the grit size was from 13 to 32 μm, covering much of the whole range of the samples investigated (6 to 36μm). The percent of grains above 120 μm was in the range 0.3 to 12% for the cooking pots whereas the entire range of all samples was 0 to 13%. Thus it seems that there was no specific different method used to obtain the grain size distribution needed to make cooking pots compared to ceramics of other uses, i.e. no addition of larger grits to change the grain size distribution. However there is a tendency for cooking pots to have more grits than the other types. The range in surface area of grits in the cooking pots is from 16 to 32% whereas in all samples it ranges from 6 to 36%. Of the 19 cooking pots observed, 60 percent are of type three, with a wide range of larger, added grains. It would seem that this method was more prevalent for increasing the total of the grits present in cooking ware. It seems that the cooking pots were made in different manners, using normal un-treated clay resources or by adding grits, but in general the total of grits present seems to be superior in the cooking pots.

Each of the grain size distribution types were used to make a full range of pots for storage, cooking ware, table ware. This suggests that the different methods of making paste (pure clay resource, clay and added grits either large or sieved), were used to obtain quite varied products. Hence the potters in this small Gaulo-Roman town used different methods to prepare the pastes over a long period of time to produce the same range of objects.

This short study attempts to indicate several ways that image analysis can be used to interpret gain size abundance in sherds. The size distribution relations can give information concerning the methods used in the production of paste for various types of ceramic objects.

Acknowledgments
We would like to thank Françoise Dumasy for her help and encouragement.

References

Betts, I. M. 1991. Thin section and neutron activation analysis of brick and tile from York and surrounding sites. In *Recent Developments in Ceramic Petrology*, eds. A. Middleton and I. Freestone. London: British Museum Occasional Paper 81, pp. 39 – 62.

Bouchain, I. 1992. Recherches sur la céramique d'Argentomagus. Maîtrise d'Histoire de l'Art et d'Archéologie. Paris: Université de Paris I.

Dumasy, F., Bouchain, I. and Rodet-Belaubi, I. 1997. L'évolution urbaine d'Argentomagus - Saint Marcel (Indre). *Revue d'Archéologie du Centre de la France* 36: 39-77.

Fieller, N. R. J .and Nicholson, P. T. 1991. Grain-size analysis of archaeological pottery; the use of statistical methods. In *Recent Developments in Ceramic Petrology*, eds. A. Middleton and I. Freestone. London: British Museum Occasional Paper 81, pp. 72 - 112.

Maggetti, M., Galetti, G, and Schnaumly, R. 1988. *Die Fein Keramik von Sissach-Brühl*, Berichte aus der Arbeit Amtes für Museen und Archäologie des Kanton Baselland, H 103.

Mason, R. B. 1995. Criteria for petrographic characterization of stone paste ceramics. *Archaeometry* 37: 307-321.

Matthew, A. J., Woods, A. J., and Oliver C. 1991. Spots

before your eyes: new comparison charts for visual percentage estimation in archaeological materials In *Recent Developments in Ceramic Petrology*, eds. A. Middleton and I. Freestone. London: British Museum Occasional Paper 81, pp. 211-276.

McLean, M. 1995. *Sedimentology*. Oxford: Oxford University Press.

Miksa, B. 1998. *A Model for Assigning Temper Provenance to Archaeological Ceramics with Case Studies from the American Southwest*. Dissertation thesis. Tucson: Department of Geosciences, University of Arizona.

Moreau, E, Sardini, P., Touchard, G. and Velde, B. 1996. 2D and 3D morphological and topological analysis of clay soil. *Microscope. Microanalysis. Microstructure*. 7: 499-504.

Parkin, T. B. and Robinson, J. A. 1992. Analysis of lognormal data. In *Advances in Soil Science*, ed. B. A. Stewart, Vol 20. Berlin: Springer Verlag, pp. 130 – 145.

Rock, N. M. S. 1988. *Numerical Geology*. Berling: Springer Verlag, see pp. 86 – 101.

Velde, B. and Druc, I. 1999. *Archaeological Ceramic Materials*. Berlin: Springer Verlag.

Whitbread, I. K. 1991. Image and data processing in ceramic petrology. In *Recent Developments in Ceramic Petrology*, eds. A. Middleton and I. Freestone. London: British Museum Occasional Paper 81, pp. 369 - 388.

Woods, A. J. 1985. Form and fabric and function: some observations on cooking pots in antiquity. In *Ceramics and Civilization, vol II Technology and Style*, ed. W. D. Kingery. Colombus, Ohio: American Ceramic Society, pp. 157 – 172.

CHAPTER 7

CRITERIA FOR EVALUATING MULTIPLE COMPONENTS IN POTTERY PASTE

Elizabeth J. Miksa
Desert Archaeology, Inc.

Prehistoric Hohokam potters of the Sonoran Desert in central Arizona are known for their red-on-buff pottery. Often, this pottery was manufactured from alluvial clays, to which caliche, salts, and crushed schist temper were added. Detailed petrographic analysis of hundreds of sherds has led to identification of several different components of red-on-buff pastes. Clay, crushed schist, and caliche are components of nearly all buff ware pottery. In addition, untempered clay lumps, silt, and sand are frequently seen. Finally, grog is present in a small proportion of the pottery. Identification of different paste components led to the observation that the primary added tempering agent, crushed schist, is compositionally much different from the grains making up the silt- to fine-sand-sized fraction of these sherds. Grain size and compositional evidence suggests that the silt- to fine-sand-sized grains are natural non-plastic inclusions, present in the clay when it was collected for pottery production. Because the different components affect final paste composition, provenance studies cannot proceed until the sources of different paste components are identified. Generalized rules for distinguishing natural from added temper have been developed through documentation of this pottery. The rules will be described, as will their application to buff ware sherds and to the problems of provenance identification.

INTRODUCTION

Ethnographers, anthropologists, and archaeo-ceramicists have long noted the complex nature of ceramic production, from aspects of raw materials procurement through manufacture and distribution to evidence of use (Arnold 1985; Bey and Pool 1992; Costin 1991; Rice 1987; Rye 1981; Shepard 1995). Chemical and petrographic analysis are well known as means of establishing provenance and production technology. In the analysis described here, petrographic analysis was initially intended as a means of establishing provenance. However, the complexity of the mixture used to create Hohokam buff ware pottery necessitated the use of petrographic analysis primarily as a tool to describe the paste. This paper presents a discussion of the qualitative methods used to characterize Hohokam buff wares in preparation for provenance analysis.

Terminology

There has been a great deal of discussion in the literature over the proper terms that should be used to describe pottery production and its associated materials. To avoid confusion, I present definitions for the terms I will use in the following discussion:

Clay mineral: A hydrous aluminum phyllosilicate mineral.
Clay: A sedimentary deposit dominated by clay-sized grains (grains less than or equal to .002 mm) used as the basis for a ceramic. Most clays contain clay minerals as well as framework silicates in the silt to sand size range.
Temper: Materials added to a clay to improve its properties for ceramic production. In most cases, temper is an intentional addition to a clay; however, some clays may be chosen because they naturally contain material that temper the clay minerals to produce a raw material with the appropriate properties. Temper may be aplastic (e.g., framework silicates) or plastic (i.e., caliche, clay minerals).
Clay body: The unfired prepared clay and temper mixture.
Paste: The fired clay and temper mixture.
Sand: Transported sediment in the .0625 to 2.0 mm size range.
Silt: Transported sediment in the .002 to .0625 mm size range.

Archaeological background

The name Hohokam is given to prehistoric peoples who inhabited central and southern Arizona in the United States from approximately AD 400 to AD 1400 (Figure 1). The Hohokam resided in semi-subterranean structures organized into settlements from the size of hamlets to large villages. They practiced irrigation agriculture of such crops as corn, beans, and squash, along with limited hunting, collection of wild plant species and agave horticulture. The Hohokam had very elaborate funerary practices and built impressive ball courts and platform mounds. Items as varied as exotic trade goods from Mexico, turquoise for jewelry, obsidian for stone tools, and regionally produced pottery were widely distributed throughout the Hohokam area. Hohokam public architecture and extensive trade networks have been interpreted as evidence of a regional network, dubbed the Hohokam Regional System (Wilcox 1979, 1980).

Buff ware pottery production and distribution was an important element of the Hohokam Regional System. Buff ware pottery was made from a light-colored clay to which caliche, salts, and crushed mica schist were added. (Caliche is micritic calcium carbonate precipitated in soil.) Geometric and zoomorphic designs were painted with iron-rich pigment, and firing was controlled enough to produce a final product with striking red or reddish-brown designs on a light buff-colored background (Plate 3a).

At one time, archaeologists thought that all, or nearly all, buff ware was made in the center of the Hohokam geographic area and distributed outward from there. Archaeological evidence of pottery production in the form of pottery anvils, clay mixing pits, and tempered and untempered clay, was found at Snaketown, one of the earliest excavated Hohokam sites. Schist quarries were found nearby (Rafferty 1982; Walsh-Anduze 1993). However, recent petrographic and chemical analyses have shown that several different production centers must have existed to account for variations in clay and temper composition. (Abbott 2000; Abbott [ed.] 2000; Abbott and Miksa 2000).

Unfortunately, the paste mixtures used for the production of buff wares do not lend themselves easily to provenance analysis. Clay composition analysis through microprobe has established that several clay compositional groups exist, but currently there is not sufficient evidence to establish unequivocally the provenance of the clays.

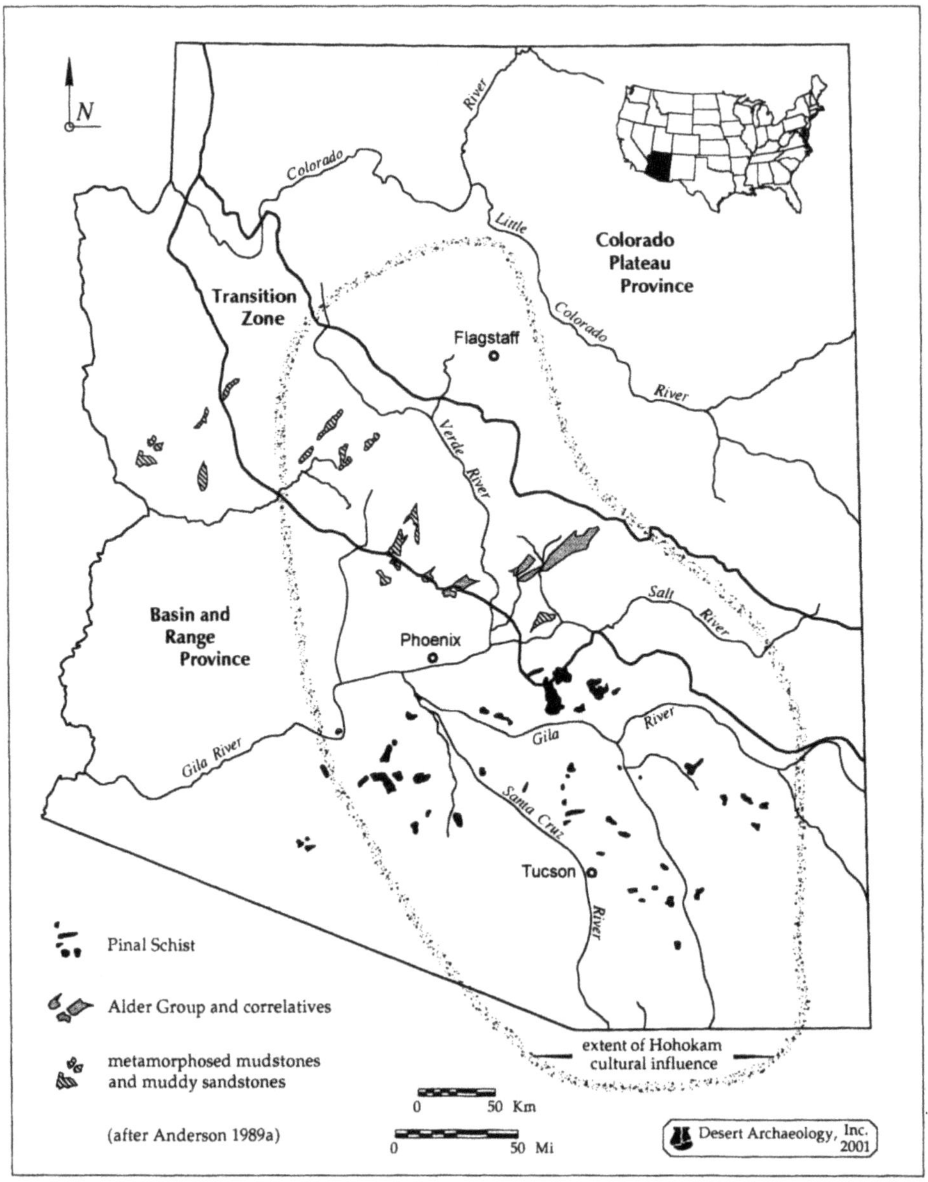

Figure 1. Location map showing the Hohokam Regional System in Arizona and the distribution of schist outcrops throughout the region.

Buff ware pastes are complex (Plate 3b). Sometimes the clay is tempered with crushed schist, sometimes with a mixture of stream sand and crushed schist. In addition, the paste contains plastic tempering material. Most buff wares contain intentionally added caliche, while some contain clay lumps that could be either the byproduct of incomplete mixing of the clay body or mixing of more than one clay. Finally, experimental work in the reproduction of buff wares suggests that salts were a required component of the pottery, either as a natural component of the clay, or added to the clay body (Abbott and Love 2000; Weisman 1987).

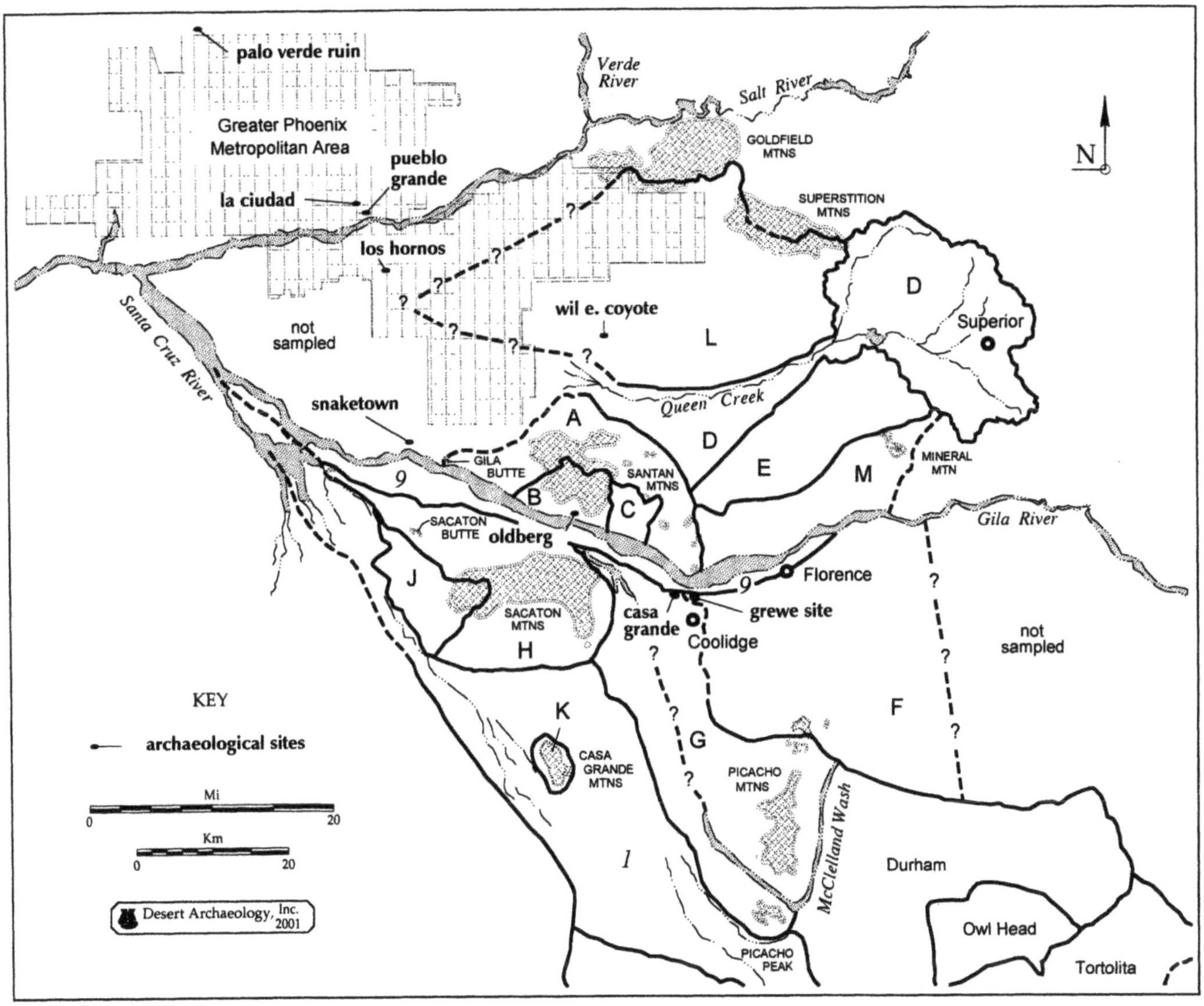

Figure 2. Petrofacies map showing the petrofacies in the study area and sites from which analyzed buff ware sherds were excavated

Geological Background

The part of Arizona that was occupied by the Hohokam is a geologically diverse area, with linear mountain ranges separated by deeply filled sedimentary basins (Menges and Pearthree 1989). Many of the mountain ranges have distinct lithologies, allowing sands derived along each range to be identified both quantitatively and qualitatively. Collection of more than 1000 sand samples throughout central and southern Arizona has led to the identification of approximately 100 petrofacies, or sand composition zones (Figure 2 shows petrofacies for the middle Gila River valley). The petrofacies are identified through a detailed process of geological map study, collection of sand samples, petrographic modal analysis (point counting) and multivariate statistical analyses (Heidke and Miksa 2000b; Miksa and Heidke 2001).

Rock types in the mountain ranges of Arizona are quite varied, ranging from Proterozoic granites, gneisses, and schists through Paleozoic sedimentary rocks, to Mesozoic and Cenozoic sedimentary and volcanic rocks. Some lithologic units are extremely restricted (less than 10 km in greatest extent) and the distance from mountain tops to valley floors is on the order of tens of kilometers. The resulting very coarse to medium subangular sands easily retain the signature of their parent bedrock. Generally, the sands within each valley or basin are distinguishable from one another; however, there is compositional overlap from basin to basin, i.e., the granitic petrofacies in one basin may be hard to distinguish from that in the next. Despite this problem, it has been possible to apply knowledge of the sand compositions in each basin to tens of thousands of pieces of sand-tempered pottery, so that the dynamics of pottery production and consumption are relatively well understood for this part of Arizona (Lombard 1987; Heidke 2001; Heidke and Miksa 2000a, 2000b; Heidke et al. 1996; Miksa 1995; Miksa and Heidke 2001).

Unfortunately, this firm identification of provenance does not extend to Hohokam pottery tempered with crushed mica schist, a rock type used primarily in the central part of the Hohokam region. The landscape exposed in this area is very old (Anderson 1989a; Dickinson 1989; Menges and Pearthree 1989). Mountain ranges are deeply incised, and several large drainages converge. Surficial deposits are dominated by rounded medium sand to silt-sized grains that do not serve well as aplastic temper. Mica schist suitable for temper is found throughout the area, exposed as small outcrops within larger mountain ranges, or as small, isolated buttes. The mica schist outcrops are the metamorphosed remnants of a Proterozoic submarine fan that was widely distributed across central and southern Arizona (Figure 1; Anderson 1989a, 1989b). The mineralogical composition of the mica schist is homogeneous over a large area. Thus, it cannot provide petrographic clues to provenance (Miksa 2000a). Provenance identification of the schist is possible through chemical analysis, but we have not yet successfully analyzed the sparse schist temper in buff wares separately from the surrounding paste. The problems in doing this relate to our inability to remove intact schist from the paste, and to the need to conduct trace element analyses in order to distinguish among schist sources. Major elements are not sufficient to make the distinction (Miksa 1998, 2000a; Ownby et al. 2001). However, increasingly precise instruments and techniques may soon make this possible.

Many buff wares also contain silt- and sand-sized materials that are not derived from the mica schist. However, it is not necessarily possible to establish the provenance of this material using the sand petrofacies models. If the temper is a mixture of materials, it is necessary to qualitatively and quantitatively separate those materials prior to submitting the sand data for classification using the petrofacies model. For instance, a clay might contain silt- to fine-sand-sized quartz and feldspar from a granitic source, and be mixed with a mica schist, plus sand from a volcanic source. A point count would establish a modal composition consistent with a mixed metamorphic-volcanic sand source, but this would be incorrect. It is necessary to establish which paste components are present prior to analysis, and to postulate their probable origin. The remainder of this paper sets out the methods and criteria that I have been using to distinguish among paste components. This involves discriminating added sediment from that which is naturally occurring, and distinguishing sand from crushed rock. Once presented, the criteria will be illustrated with examples.

PETROGRAPHIC METHODS

Qualitative description of ceramics has a long history in and of itself (e.g., Middleton and Freestone [Eds.] 1991; Peacock 1967; Peacock 1977; Peacock and Williams 1997; Shepard 1936, 1942; Stoltman 1989), and the criteria presented here are not new. However, using qualitative criteria explicitly to identify paste components for quantitative analysis is not often done. Hence, it is worthwhile to consider the criteria that are used in this process, and the ways in which they are applied.

For sand-tempered Hohokam pottery, the primary goal of petrographic analysis is to establish the provenance of the sand. In multiple-component buff ware pottery, composition may reflect several provenance signatures. A valid interpretation of these signatures requires them to be quantified separately. To illustrate this problem, I will briefly describe the point count method most commonly employed for pottery in Arizona, followed by the criteria used to identify buff ware paste components. The final sections illustrate how this information is applied to the point count data.

Petrographic modal analysis

Petrographic modal analysis, or point counting, is an established method of quantifying the mineralogical composition of a material such as a rock or a sand (Chayes 1956). To point count, one establishes a grid over a sample, then counts the composition of the material under each grid point. The categories counted depend on the research problem(s) under study and the types of materials. For instance, to gain a basic understanding of a rock, one might count only three categories, known as *grain parameters*: Quartz, Feldspar, and Lithic fragments. In Arizona, there is a great deal of compositional variability in the bedrock, such that it is possible to make detailed distinctions between rocks and/or the sediments derived from them. Table 1 (pp. 92-93) presents one such scheme used to distinguish among monocrystalline grains and rock fragments. In this scheme, 43 rock and mineral parameters are used.

One problem that must be resolved in point counting is that of whether to count coarse-grained rocks such as granite, as a rock type or as the individual minerals that make up the rock. Since coarse-grained rocks may break down into their constituent minerals, it is most appropriate to count them as such when creating petrofacies models (Figure 3). This "Gazzi-Dickinson" point counting method (Dickinson 1970; Gazzi 1966) emphasizes the mineralogical composition in order to minimize maturity or transport effects from rock to sediment. It allows statistical comparison of sediments throughout a basin, provided they are of roughly the same stream order (Ingersoll et al. 1993).

Characterizing buff ware pastes

Although petrographic modal analysis has been used primarily as a provenance tool, the compositional characterization of sand sized grains can also be used to quantify the proportion of different components in a paste. Table 1 shows how parameters can be defined to count the grains derived from schist separately from those derived from other rock types. In this case the distinction is made on the basis of clearly identifiable petrographic criteria; for instance, SQTZ is counted rather than quartz, when a grain is in a schist fragment or clearly has attributes associated with quartz originating in the schist (attached mica, etc.). This type of point counting is independent of the origin of the schist-it could be from sand, naturally occurring in the clay, or intentionally crushed and added to the clay body as temper. The interpretation of the genetic origin of the schist is based on qualitative criteria.

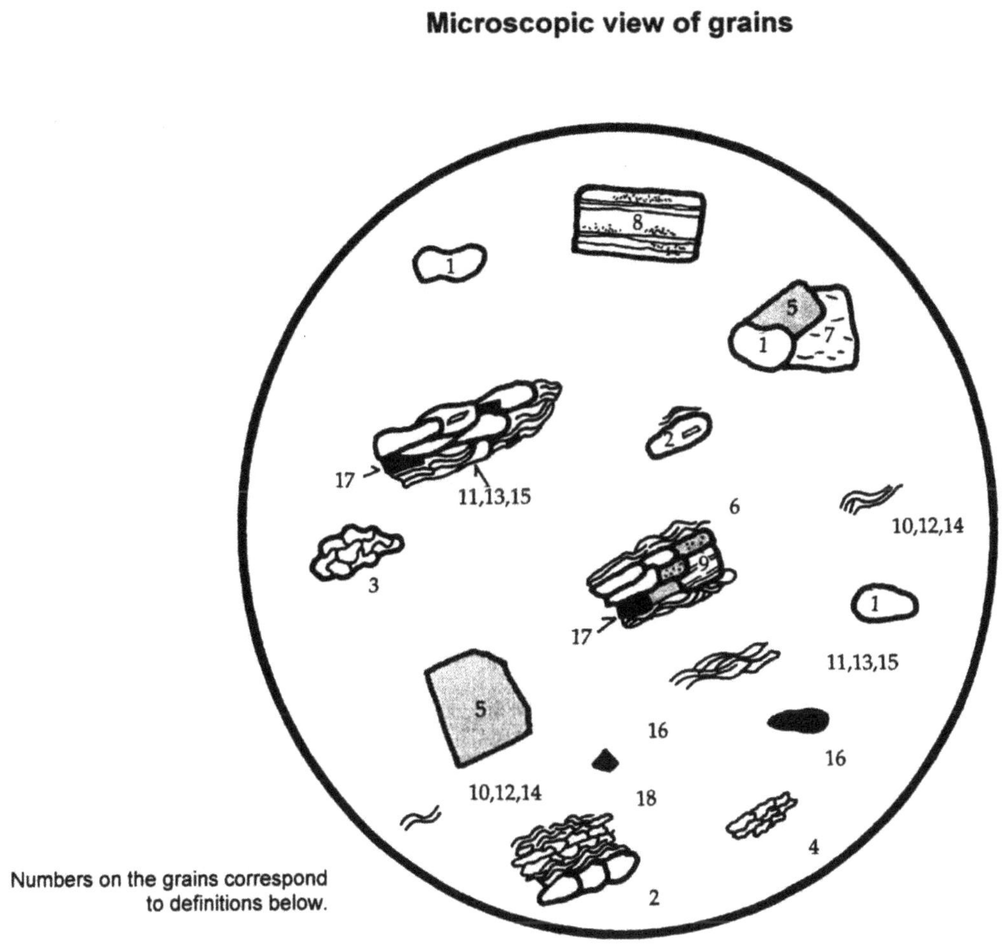

Qtz (1) Grains of quartz, either alone or in rock fragments.

SQtz (2) Sand-sized grains of quartz contained within schist[a] or with features showing that they come from schist.

LMM (3) Metamorphic aggregates: Metamorphosed quartz crystals in which the individual crystals are smaller than sand-sized. Although the quartz crystals have crenulated boundaries, they are not foliated.

LMF (4) Foliated metamorphic aggregates: Metamorphosed quartz crystals in which the individual crystals are smaller than sand-sized. The overall fabric of these grains is foliated.

Kspar (5) Potassium feldspar grains that do not originate in schist or gneiss.

Skspar (6) Potassium feldspar grains contained within schist or with features showing that they come from schist.

Plag (7) Plagioclase feldspar, stained pink, in which alteration and/or sericitization affect less than 10% of the grain.

Plagal (8) Altered: Plagioclase feldspar crystals with 10% to 90% of the grain disrupted by the growth of new crystals. Alteration products include sericite, clay minerals, carbonate, and epidote.

SPlag (9) Plagioclase feldspar crystals contained within schist or with features showing that they come from schist. No distinction is made with regard to alteration state.

Musc[b] (10) and Smusc (11) Muscovite grains that do not originate from schist, and those that do, respectively.

Biot[b] (12) and Sbiot (13) Biotite grains that do not originate from schist, and those that do, respectively.

Chlor[b] (14) and Schlor (15) Chlorite grains that do not originate from schist, and those that do, respectively.

Opaq (16) and Sopaq (17): Opaque minerals that do not originate from schist, and those that do originate from schist, respectively.

LMT (18) A schist fragment comprising grains that are finer than sand-sized. Note that a single shist fragment might have some areas of fine grains, and other areas with sand-sized grains, due to the typical banding seen in schists.

[a]The term "schist" is used here, but this counting method applies to any coarse foliated rock type, not just schist.
[b]Mica grains are not illustrated separately. One or more types of mica may occur in any given foliated rock.

Figure 3. Illustration of the Gazzi-Dickinson point counting method.

Distinguishing Crushed Rock from Sand

To separate crushed schist from that which occurs naturally, I examined and described natural schist-bearing sands (middle Gila River Petrofacies "E" in Figure 2) as an example of a non-crushed source of sand-sized foliated rock fragments (Plate 3c). I also crushed pieces of mica schist and made "briquettes", experimental sherds that were made by combining the crushed schist with commercial clay (Plate3d). This allowed me to develop criteria for identifying crushed rock temper in sherds from the Grewe site in central Arizona (Miksa 2000a). This was followed by observation of other schist-rich sands and of crushed-schist-tempered sherds (Abbott [ed.] 2000; Miksa 2000b). The criteria listed here may not be exactly the same as criteria needed for other studies, but they serve as examples of the types of differences that can be used to distinguish among aplastic temper types.

Grain Size Distribution

In crushed schist tempers, the schist fragments tend to be very coarse sand to granule size. Although there are smaller grains of schist, the coarse grains are visually and volumetrically dominant. The schist also breaks down into its constituent fine to medium sand-sized grains of quartz, feldspar, and micas. Where non-schist grains are present, they tend to be fine to medium sand size as well. The combination of the very coarse sand-sized grains of schist and the fine to medium sand-sized grains of disaggregated schist or sand leads to a strongly bimodal grain size distribution (Plate 3d). In the unmodified schist-bearing sands, the schist fragments tend to be the same size as other grains in the sand; the grain size distribution is unimodal (Plate 3c).

Rock Homogeneity

For crushed schist tempers, the texture of the schist in any given briquette or sherd is homogeneous, with a limited range of textures and compositions (Plate 3d). In the schist-bearing sands, there is a wider range of schist textures and compositions because many different outcrops contribute fragments to each stream (Plate 3c).

Grain Angularity

Because of the platy habit of schist, grains that have been transported in streams do not become strongly rounded. Thus, the crushed schist grains are not noticeably angular compared to stream transported schist grains (Plate 3c-d). There is an increase in angularity in the crushed grains, but the increase is very subtle and difficult to evaluate with a standard petrographic microscope. Video imaging software might be able to measure this difference, but qualitative estimates of the range and abundance of grains in different angularity classes are not sufficient to demonstrate this difference.

Crushed Grain Features

Some crushed schist grains have conchoidally fractured quartz and frayed mica grains, attributes visible in thin section and verified through SEM photography (Ownby et al. 2001; Plate 3d). These features are not uniformly present, and they can occur on transported grains as well, so their presence is not definitive. Like the angularity of grains, this is a subtle difference of proportions of attributes throughout a thin section, and should be used with caution when distinguishing among possible temper types.

Finely Disseminated Mica

In the experimental briquettes (Plate 3d), there were tiny grains of mica disseminated throughout the paste. In the process of crushing schist for the experimental briquettes, it was noted that the micas were readily reduced to clay-to-silt-sized grains. Nearly all schist-tempered sherds have pastes rich in clay to silt sized mica. Many of these also have clay lumps that are free of mica. Thus, the presence of finely disseminated micas was accepted as an indication of the addition of crushed schist to a clay.

Large Volcanic Rock or Quartzite Fragments. In the experimental briquettes, it was common to find a few grains of volcanic rock from the implements used to crush the schist (Plate 3d). In schist-tempered sherds, it is not uncommon to find a few grains of volcanic rock or quartzite that are incompatible with the composition of other temper grains. However, their presence or absence alone is not definitive evidence for determining temper type (Miksa 2000b).

The above criteria must be used together in interpreting the origin of temper in a paste. For instance, the presence of frayed micas and fractured quartz, in the absence of other features associated with known crushed schist tempers, would not be enough in and of itself to lead to an interpretation of crushed schist temper.

Observations of Sand temper

Establishing the origin of sand temper in the buff wares is more difficult than working with crushed schist. Where sand is present, it tends to make up a small proportion of the paste. There are fewer grains than would be optimal for the evaluation of the origin of the sand. One clue to the origin of the sand lies in its relationship to the silt-sized grains in the paste. If the sand texture and composition is continuous with that of the silt, it seems more likely that it occurs as a coarser end member of the clay collected for pottery production than as an intentional addition. This is not definitive proof. Sand and clay from the same parent material could also show these features. However, intentional sand temper in Hohokam pottery tends to be abundant and coarse. Thus, the following criteria are relevant to the problem:

Grain size distribution

In sand-tempered Hohokam pottery the sand grains tend to be medium to very coarse sand-sized. When a silt to fine sand fraction is present it forms a second mode in the grain size distribution. In sherds in which sand is postulated to have been part of the clay, there is a unimodal grain size distribution from coarse silt to medium sand. Plate 3e is a photomicrograph illustrating a unimodal grain size distribution in a buff ware, while Plate 3f illustrates a typical bimodal grain distribution.

Mineralogical composition, including textural characteristics of grains

There may be compositional continuity between the silt and sand fractions, meaning that the same grain types occur in the same frequencies in each fraction, or that the compositional changes by size fraction are regular, and can be reasonably attributed to weathering or erosional processes. The photomicrographs of buff ware in Plate 4a-b illustrate such continuity. On the other hand, there may be a compositional break between the silt and sand fractions, or a change in composition that cannot be reasonably attributed to weathering or erosion. Plates 4c-4d illustrate such variation.

The variation between silt and sand fractions can be true not only for overall composition but for the attributes of the grains of each composition. For instance, quartz in the silt fraction could show features indicative of one origin, for instance clear, unstrained bipyramidal grains from a volcanic rock, while the quartz in the sand fraction could show bubble trains and strain patterns typical of granitic rocks. All aspects of mineralogy and genetic origin of the silt and sand-sized grains are relevant to the question of compositional continuity between the size fractions.

Grain angularity

As in distinguishing between sand and crushed schist temper, it is possible to examine the difference in angularity between the silt and sand fractions of the paste. Features to examine are whether the angularity of the different size fractions similar or if there is a regular change in angularity with grain size. Again, this is a difficult criteria to apply, and is often inconclusive. It is possible that visual imagery software would make it more possible to assess this attribute of the size fractions.

Interpretation of the petrographic observations

I hypothesize that in most cases, continuity between sand and silt in terms of texture and composition indicates the likelihood of them being part of the same deposit. However, a final determination must be made on the basis of examination of known source materials, and with reference to the types of materials used in pottery manufacture as a whole. All of these criteria are bound by the context in which they occur, and the criteria will change for each study area depending on the basic materials available and the technology of manufacture employed by the artisans who made the pottery. Without basic knowledge of these contexts no interpretation can be made.

EXAMPLES

To illustrate the importance of understanding the context and materials used in pottery manufacture, I present two examples of known cases, and then discuss the implications of this knowledge for the Hohokam buff wares. The two examples are chosen to illustrate the points made above. The first example is from the Philippines–clearly far from the Hohokam sphere of influence! However, the clay used by potters in the Philippine example is very similar to clays used in parts of Arizona. Furthermore, the ethnographic nature of the study provides a fully "known" context in which to illustrate full continuity between size fractions in a paste. The second example comes from the Hohokam region. Although the pottery in this example is prehistoric, much is known about its procurement and production; thus providing a secure context for evaluating the paste. In this example, there is compositional continuity between the size fractions, but a break in grain size distribution. These two known examples are then followed by application of the illustrated principles to the "unknown" case of the Hohokam buff ware sherds.

Philippines

An ethnoarchaeological study carried out by Miriam Stark recovered pots and clays from the Pasil River valley, in northern Luzon, Philippines (Stark 1993). Stark spent several field seasons talking to potters, learning their favorite clay sources, and commissioning vessels. She collected well over 200 pots made by known potters from known sources, as well as several raw clay samples that were prepared by local potters but not made into vessels. Pasil potters used the clays without additions, though they did prepare their materials by removing large grains and by pounding the remaining material to achieve a uniform texture (Stark 1993). The pounding was not sufficiently vigorous to affect sand-sized grains.

Fired pots were prepared for chemical and petrographic analysis by removing 2.5 cm disks from their bases (Stark et al. 2000). Briquettes were made from the clay samples for comparable analyses. Subsequent neutron activation and petrographic modal analyses were able to establish compositional groups for this known set of pottery and clay source materials.

The clays used by Pasil potters are derived from Tertiary age intermediate volcanic materials and have abundant silt- to sand-sized grains. All of the clays included plagioclase and hornblende. Clays from most of the sampled localities include minor volcanic rock fragments, though two clay localities had more abundant volcanic rock fragments (Miksa 1997; Stark et al. 2000). Interestingly, the minerals included in the clay show little sign of weathering. Photomicrographs of the clays (Plate 4e-f) clearly show the continuity of grain size, texture, and composition between the sand and silt fractions of these clays. In this case, the continuity is known to be a result of the use of a sandy clay, in which grains were removed prior to pottery production, rather than added.

Tucson

Prehistoric potters in the Tucson area also employed a clay from a volcanic parent material for their pottery. In this case, however, the clay has silt-sized grains in the parent material but few sand-sized grains. No ethnographic evidence is available, but clay source locations are well known, and are used by modern potters in the area to manufacture Hohokam-like pots. In addition, nearby prehistoric pottery workshops have been excavated, some of which contained clay and temper in various stages of production (Huntington 1986; Henry Wallace, personal communication, November 2000). Temper is derived from nearby sands draining the same volcanic deposits as those on which the clays formed, thus there is compositional continuity between the silt and sand fractions (Plate 4g-h). However, even with compositional continuity, it is possible to identify an added sand component: The sand temper in this sherd is more abundant than in the Pasil example. It is also more rounded, suggesting transport. In addition, the grain size distribution from the silt to sand-size fractions is bimodal rather than unimodal.

Characterizing Hohokam Buff Ware pastes

The previous examples suggest that the criteria developed above can be applied to the plastic tempers in complex

Hohokam buff ware pastes, as long as they are used with caution and in context. The final problem is to assess the aplastic component.

Two major aplastic temper components are visible in buff wares. The first is caliche, visible as large rounded micritic calcium carbonate grains. These large grains are not visible in raw materials. Experimentation suggests that they were added to help achieve the light buff color of the clay (Abbott and Love 2000; Weisman 1987).

The second aplastic component is expressed as "lumps" or areas of clay that are free, or nearly free, of silt- to sand-sized grains, including finely disseminated mica (Plate 3g). It is possible that these indicate the mixing of two or more clays; however, the clays in the lumps are indistinguishable from those in the rest of the paste. Furthermore, the alluvial clays apparently used by the Hohokam for buff wares are extremely variable. Plate 3h is an example of early pottery (ca. AD 400) from the Tucson area (Heidke 1999). It is barely fired enough to be called a ceramic, and the material has no added temper. This clay has the same composition and textures as clay available in the Santa Cruz River. Note the clay lumps and the variation in silt content across just the width of this thin section. It is probable that the buff ware potters were using similar alluvial clay (Weisman 1987), with included silts and clay lumps.

Applying the criteria

The criteria detailed in this paper have been applied to 130 thin sections of buff ware pottery from nine site areas throughout central Arizona (Figure 2). Unfortunately, many thin sections were counted before all of the categories had been recognized and quantified; these do not have all classes of data. However, data on crushed schist, sand, clay lumps, and caliche components are available for 69 thin sections. In addition, sand data for these thin sections have been submitted for provenance analysis. Finally, most of these sherds have been analyzed chemically by microprobe. Results of this analysis indicate that different clays were used (Abbott 2000), but the provenance information on clays is not precise enough to solve the archaeological problem of trade within the Hohokam world. During point counting, I noted that there were several paste "recipes", that is, relatively standard paste mixtures seen in many of the thin sections. This is a qualitative observation, based on a number of thin sections that "look like" one another overall, in terms of color, texture, and composition. Prior to data analysis, I estimated that there were approximately five major compositions, and I hypothesize that these might represent major buff ware producers in the Hohokam regional system.

The paste composition data do not fully support the visual observation. They have been evaluated using correspondence analysis and K-means cluster analysis. The correspondence analysis, used as a means of exploratory data analysis, was inconclusive. The K-means cluster analysis suggests that five to seven major paste groups are represented. Figure 4 is a stacked bar graph illustrating the paste composition of the five cluster solution. It is obvious that there are differences between the groups, but the differences are not strong. This may mean that the groups are very similar because of the cultural constraints on buff ware production, or it may mean that it is difficult to quantify the visible differences seen with the eye, or it may mean that the differences do not really exist.

It seems likely that differences do exist; however, since chemical data support the existence of several paste groups (Abbott 2000). One way to test the strength of paste clusters is to evaluate the sand composition within the groups. If the clusters correspond to provenance, then the sand compositions within each cluster should be similar. A preliminary test of this assumption shows that the correlation between cluster group and provenance is not significant. The preliminary test was done by comparing the petrofacies membership of the sand fraction of each sherd predicted by the set of nested discriminant models that have been developed for the middle Gila River valley (Miksa 2000a, 2000b) with the cluster membership given by the five cluster model. The results of this analysis are shown in Tables 2a and 2b. Note that there are two levels of the discriminant model shown here (Heidke and Miksa 2000b). The first level divides the tempers into "Lithic-rich" versus "Mineral-rich" sands, the second shows the division of "Mineral-rich" tempers into individual petrofacies.

Table 2. Cross-tabulation of cluster assignments with predicted petrofacies membership for the sand fraction of buff ware tempers. See Figure 2 for petrofacies locations.

a. Generic-level assignments

	Generic sand groups		
Cluster groups	Lithic-rich (D,E,F,G,K,L,M)	Mineral-rich (A,B,C,H,J)	Total
1	5	13	18
2	7	10	17
3	2	6	8
4	2	12	14
5	4	8	12
Totals	20	49	69

Chi-squared=2.882, df=4, prob.=0.578.

b. Predicted petrofacies membership for sherds assigned to the mineral-rich group.

	Predicted petrofacies membership					
Cluster groups	A	B	C	H	J	Total
1	0	11	0	2	0	13
2	1	3	2	1	3	10
3	1	5	0	0	0	6
4	0	8	0	0	4	12
5	1	4	0	1	2	8
Totals	3	31	2	4	9	49

Chi-squared=22.916, df=16, prob.=.116

Detailed examination of the point count data for the sand fraction shows that there are significant differences between the composition of the middle Gila River sands and the sand fraction of the sherds. For instance, the metamorphic content of the sand fraction of the temper is higher than the natural sands by a factor of four. This seems to be due in part to an inability to fully separate the schist fraction of the temper from the sand. To avoid bias when point counting, I assume that a grain belongs to the sand fraction unless it shows definite features to the contrary. This overinflates the sand fraction to some degree, but provides necessary consistency to the count. The discordance between sand and sand tempers will need to be evaluated and explained before provenance work can continue. Be that as it may, the problems with the provenance data do not affect the preliminary conclusion that there are different paste recipes (Figure 4).

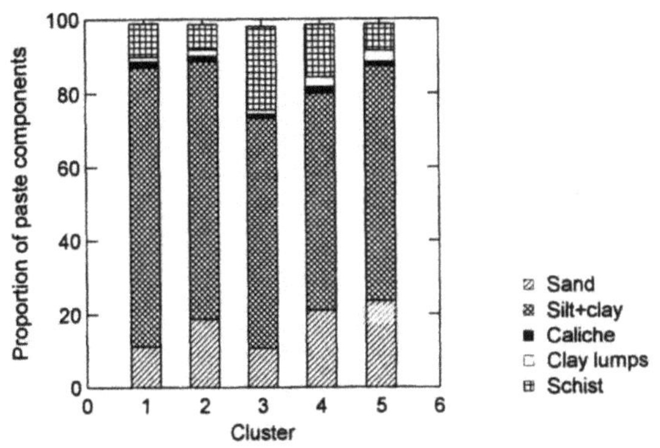

Figure 4. Paste composition for the samples included in the five group K-means cluster analysis.

CONCLUSIONS

Clearly, there are many problems to be solved before we can characterize the provenance of Hohokam buff wares. The descriptive criteria set forth in this paper are a first step in characterizing a complex paste in a way that can lead to quantification, and subsequent hypothesis testing through statistical means. The data presented here demonstrate the importance of separating out the various paste components before any statistical analyses can proceed. The data also emphasize the difficulty of documenting the complexities that can be seen with the human eye and understood through long experience: The reduction of data to a set of statistical equations rarely accounts for the subtleties processed by the human brain. In the case of Hohokam buff ware, I hope that further analysis and documentation will lead either to firmer confirmation of the paste recipe hypothesis, or to a strong refutation.

Acknowledgments

I received help from many people in putting these ideas on paper, and without them I would not have been able to complete this chapter. Dave Abbott keeps me involved with buff wares and provided information about buff ware chemistry. Jim Heidke keeps me on the path of logic in data analysis. Dave Killick provided the photomicrograph facilities; Rob Ciaccio and Catherine Gilman drafted the figures; Emilee Mead, Scott van Keuren, Andrea Matthews, and Linda Friedman all provided assistance with graphics. Owen Lindauer allowed me to use his photo of the buff ware bowl (Plate 3.a). Finally, I thank Isabelle Druc for her patience and extensive editorial assistance. Any mistakes remaining in this document are entirely my own.

References

Abbott, D.R. 2000. Hohokam buffware production and distribution and the provenance of decorated pottery for the BMGR. In *Procurement, processing, and passing through*, ed. J.D. Lyon. Cultural Resource Report No. 98-267. Tucson, Arizona: SWCA, pp. 577-620.

Abbott, D.R. ed. 2000. *The Grewe Archaeological Research Project, Volume 2: Ceramic studies*. Anthropological Papers No. 99-1. Tempe, Arizona: Northland Research (in press).

Abbott, D.R. and Love, T.K. 2000. Specialized studies and observations on pottery production. In *The Grewe Archaeological Research Project, Volume 2: Ceramic Studies*, ed. D. R. Abbott. Anthropological Papers No. 99-1. Tempe, Arizona: Northland Research, in press.

Abbott, D.R. and Miksa, E.J. 2000. Progress on buffware provenance. Paper presented at the Spring 2000 meeting of the Arizona Archaeological Council, Coolidge, Arizona.

Arnold, D.E. 1985. *Ceramic theory and cultural process.* Cambridge: Cambridge University Press.

Anderson, Phillip 1989a. Proterozoic plate tectonic evolution of Arizona. In *Geologic Evolution of Arizona*, eds. J. P. Penney and S. J. Reynolds. Digest 17. Tucson, Arizona: Arizona Geological Society, pp. 17-56.

Anderson, Phillip 1989b. Stratigraphic Framework, Volcanic-Plutonic Evolution, and Vertical Deformation of the Proterozoic Volcanic Belts of Central Arizona. In *Geologic Evolution of Arizona*, eds. J. P. Penney and S. J. Reynolds. Digest 17. Tucson, Arizona: Arizona Geological Society, pp. 57-148.

Bey, G.J., III, and Pool, C.A. eds. 1992. *Ceramic Production and distribution: An integrated approach.* Boulder: Westview Press.

Chayes, F. 1956. *Petrographic modal analysis*. New York: John Wiley and Sons.

Costin, C.L. 1991. Craft specialization: issues in defining, documenting, and explaining the organization of production. In *Archaeological method and theory, Vol. 3*, ed. M.B. Schiffer. Tucson: University of Arizona Press, pp. 1-56.

Dickinson, W.R. 1970. Interpreting detrital modes of graywacke and arkose. *Journal of Sedimentary Petrology* 40, 695-707.

Dickinson, W.R. 1989. Tectonic setting of Arizona through geologic time. In *Geologic evolution of Arizona*, eds. J.P. Jenney and S.J. Reynolds. Digest 17. Tucson, AZ: Arizona Geological Society, pp. 1-16.

Gazzi, P. 1966. Le arenarie del flysch sopracretaceo dell'Alppeninno modenese: correlazioni con il flysch di Monghidoro. *Mineralogica e Petrografica Acta* 12, 69-97.

Heidke, J.M. 1999. Cienega phase incipient plain ware from southeastern Arizona. *Kiva* 64(3), 311-338.

Heidke, J.M. 2001. Utilitarian ceramic production and distribution in the prehistoric Tonto Basin. In, *2000 years of settlement in the Tonto Basin: Tonto Creek Archaeological Project overview and synthesis (draft)*, ed. J.J. Clark. Anthropological Papers No. 25. Tucson, Arizona: Center for Desert Archaeology.

Heidke, J.M., Kamilli, D.C., and Miksa, E. 1996. *Petrographic and qualitative analyses of sands and sherds from the Lower Verde valley.* Technical Report No. 95-1. Tucson, Arizona: Center for Desert Archaeology.

Heidke, J.M. and Miksa, E.J. 2000a. Ceramic temper provenance studies. In *Tonto Creek Archaeological Project, Volume 1: A Tonto Basin perspective on ceramic economy*, eds. J.M. Vint and J.M. Heidke, Anthropological Paper No. 23. Tucson, Arizona: Center for Desert Archaeology, pp. 95-146.

Heidke, J.M. and Miksa, E.J. 2000b. Correspondence and discriminant analyses of sand and sand temper compositions, Tonto Basin, Arizona. *Archaeometry* 42, 273-299.

Huntington, F.W. 1986. *Archaeological investigations at the West Branch site: Early and Middle Rincon occupation in the southern Tucson Basin.* Anthropological Papers No. 5. Tucson, Arizona: Institute for American Research.

Ingersoll, R.V., Kretchmer, A.G., and Valles, P.K. 1993. The effect of sampling scale on actualistic sandstone petrofacies. *Sedimentology* 40, 937-953.

Lombard, J.P. 1987. Provenance of sand temper in Hohokam ceramics, Arizona. *Geoarchaeology* 2, 91-119.

Menges, C.M. and Pearthree, P.A. 1989. Late Cenozoic tectonism in Arizona and its impact on regional landscape evolution. In *Geologic evolution of Arizona*, eds. J.P. Jenney and S.J. Reynolds. Digest 17. Tucson: Arizona Geological Society, pp. 649-680.

Middleton, A. and Freestone, I. (Eds.) 1991 *Recent Developments in Ceramic Petrology.* Occasional Papers No. 81, London: British Museum.

Miksa, E.J. 1995. Petrographic analysis of sherds for the McDowell to Shea Boulevard portion of the Beeline Highway excavation project. In *Archaeology at the head of the Scottsdale canal system, Volume 2: Studies of artifacts and biological remains*, eds. T.K. Henderson and D.R. Abbott. Anthropological Papers No. 95-1. Flagstaff, Arizona: Northland Research, Inc., pp. 15-38.

Miksa, E.J. 1997. Petrographic analysis of fired clays from the villages of Dalupa and Dangtalan, Kalinga, northern Luzon, Philippines. Petrographic report 97-1. Tucson, Arizona: Desert Archaeology, Inc.

Miksa, E.J. 1998. A model for assigning temper provenance to archaeological ceramics with case studies from the American Southwest. Unpublished doctoral dissertation. Tucson, AZ, University of Arizona.

Miksa, E.J. 2000a. Petrologic studies of sands, rocks, and tempers in the middle Gila River valley, Arizona. In *The Grewe Archaeological Research Project, Volume 2: Ceramic Studies*, ed. D. R. Abbott. Anthropological Papers No. 99-1. Tempe, Arizona: Northland Research, in press.

Miksa, E.J. 2000b. *Provenance characterization of schist-tempered buff ware pottery.* Petrographic Report No. 99-6. Tucson, Arizona: Desert Archaeology, Inc.

Miksa, E.J. and Heidke, J.M. 2001. It all comes out in the wash: Actualistic petrofacies modeling of temper provenance, Tonto Basin, Arizona, USA. *Geoarchaeology* 16(2):177-222.

Ownby, M.F., Ownby, C.L., and Miksa, E.J. 2001. Using scanning electron microscopy to characterize schist as a temper: An experimental exercise. Poster presented at the 102nd annual meeting of the Archaeological Institute of America meetings, January 4-6, 2001, San Diego, California.

Peacock, D.P.S. 1967. Romano-British Pottery Production in The Malvern District of Worcestershire. *Transactions of The Worcestershire Archaeological Society* 1:15-28.

Peacock, D.P.S. 1977. Roman Amphorae: Typology, Fabric, and Origins. *Ecole Francaise De Rome* 32:261-278.

Peacock, D. and Williams, D. 1997. A little bit of dirt. *Journal of Archaeological Science* 24, 1089-1091.

Rafferty, K. 1982 Hohokam micaceous schist mining and ceramic craft specialization: an example from Gila Butte, Arizona. *Anthropology* 6(1-2), 199-222.

Rice, P.M. 1987. *Pottery analysis: A sourcebook.* Chicago: University of Chicago Press.

Rye, O.S. 1981. *Pottery technology: Principles and reconstruction.* Manuals on Archaeology No. 4. Washington, D.C.: Taraxacum Press.

Shepard, A.O. 1936. Technology of Pecos pottery, In *The pottery of Pecos, Vol. 2*, eds. A.V. Kidder and A.O. Shepard. Papers of the Phillips Academy Southwestern Expedition No. 7. Andover, Massachusetts, pp. 389-587.

Shepard, A.O. 1942. *Rio Grande glaze paint ware: A study illustrating the place of ceramic technological analyses in archaeological research.* Contributions to American Anthropology and History, 7(39). Washington, D.C.: Carnegie Institution.

Shepard, A.O. 1995. *Ceramics for the archaeologist.* Publication No. 609. Washington, D.C.: The Carnegie Institution.

Stark, M.T. 1993. Pottery Economics: A Kalinga Ethnoarchaeological Study. Unpublished Ph.D. dissertation, Department of Anthropology. Tucson, Arizona: University of Arizona.

Stark, M.T.; Bishop, R.L. and Miksa, E. 2000. Ceramic technology and social boundaries: Cultural practices in Kalinga clay selection and use. *Journal of Archaeological Method and Theory* 7(4):295-331.

Stoltman, J.B.; 1989. A quantitative approach to the petrographic analysis of ceramic thin sections. *American Antiquity* 54(1):147-160.

Walsh-Anduze, M.E. 1993. The sourcing of Hohokam red-on-buff ceramics using inductively coupled plasma spectroscopy: "Schist happens." Unpublished Master's thesis, Department of Anthropology, Flagstaff, Arizona: Northern Arizona University.

Weisman, R. 1987. Pioneer to Sedentary ceramic technology at La Ciudad. In *Specialized studies in the economy, environment and culture of La Ciudad*, eds. J.E. Kisselburg, G.E. Rice, and B.L. Shears. Office of Cultural Resource Management. Tempe: Arizona State University, pp. 1-40

Wilcox, D.R. 1979. The Hohokam regional system. In *An archaeological test of sites in the Gila Butte-Santan region, south-central Arizona*, eds. G.E. Rice, D. Wilcox, K. Rafferty, and J. Schoenwetter. Anthropological Research Papers No. 18. Tempe: Arizona State University pp. 77-116.

Wilcox, D.R. 1980. The current status of the Hohokam concept. In, *Current issues in Hohokam prehistory: Proceedings of a symposium*, eds. D. Doyel and F. Plog. Anthropological Research Papers No. 23. Tempe: Arizona State University, pp. 236-242.

Table 1. Parameters used for the buff ware point counts.

Parameter	Description
Monomineralic Grains	
QTZ	All sand-sized quartz grains except those derived from or contained within coarse foliated rocks[1]. Unstained.
KSPAR	Alkali feldspars, except those derived from or contained within coarse foliated rocks. Potassium feldspar stained yellow, unstained plagioclase feldspar, perthite, antiperthite.
MICR	Microcline/anorthoclase: Alkali feldspar with polysynthetic (cross-hatch) twinning, stained yellow or unstained.
PLAG	Plagioclase feldspar, stained pink, except for grains derived from or contained within coarse foliated rocks. Grains commonly have albite twinning and/or carlsbad twinning. Alteration, sericitization affect less than 10% of the grain.
PLAGAL	Altered plagioclase, except for grains derived from or contained within coarse foliated rocks. Alteration affects 10% to 90% of the grain. Alteration products include sericite, clay minerals, carbonate, epidote.
PLAGGN	Considerably altered plagioclase, except for grains derived from or contained within coarse foliated rocks. Alteration affects more than 90% of the grain.
PX	Undifferentiated members of the pyroxene group.
AMPH	Undifferentiated members of the amphibole group.
OLIV	Olivine
OPAQ	Undifferentiated opaque minerals such as magnetite/ilmenite, rutile, and iron oxides.
MUSC	Muscovite mica.
BIOT	Biotite mica.
CHLOR	Chlorite group minerals
SPHENE	Sphene.
EPID	Undifferentiated members of the epidote family (epidote, zoisite, clinozoisite).
GAR	Undifferentiated members of the garnet group.
CACO	Calcium carbonate minerals.

Parameter	Description
Monomineralic Grains in Coarse Foliated Rocks[1]	
SQTZ	Quartz derived from or contained within coarse foliated rocks.
SPLAG	Plagioclase feldspar derived from or contained within coarse foliated rocks.
SMUSC	Muscovite mica derived from or contained within coarse foliated rocks.
SBIOT	Biotite mica derived from or contained within coarse foliated rocks.
SCHLOR	Undifferentiated chlorite group minerals derived from or contained within coarse foliated rocks.
SOPAQ	Undifferentiated opaque minerals derived from or contained within coarse foliated rocks.
SKSPAR	Potassium feldspar derived from or contained within coarse foliated rocks.
Volcanic Lithic Fragments	
LVF	Felsic volcanic such as rhyolite: Microgranular nonfelted mosaics of submicroscopic quartz and feldspars, often with microphenocrysts of feldspar, quartz or rarely ferromagnesian minerals. Groundmass is fine to glassy, always has well developed potassium feldspar (yellow) stain, may have plagioclase (pink) stain too.
LVFB	Biotite-bearing felsic volcanic: Microgranular nonfelted mosaics of submicroscopic quartz and feldspars, often with microphenocrysts of feldspar, quartz, always with phenocrysts of biotite. Groundmass is fine to glassy, always has well developed potassium feldspar (yellow) stain.
LVI	Intermediate volcanic rock such as rhyodacite, dacite, latite, and andesite.
LVM	Basic volcanic: Visible microlites or laths of feldspar crystals in random to parallel fabric, usually with glassy or devitrified or otherwise altered dark groundmass. Often with phenocrysts of opaque oxides, occasional quartz, olivine, or pyroxene. Rarely yellow stained, often very well developed pink stain, representing intermediate to basic lavas such as latite, andesite, quartz-andesite, basalt, or trachyte.
LVV	Glassy volcanics: Vitrophyric grains showing relict shards, pumiceous fabric, welding, or perlitic structures, sometimes with microphenoxrysts, representing pyroclastic or glassy volcanic rocks.
LVH	Hypabyssal volcanics (shallow igneous intrusive rocks): Equigranular anhedral to subhedral feldspar-rich rocks with no glassy or devitrified groundmass, coarser-grained than LVF, most have yellow and pink stain.

Parameter	Description
Sedimentary Lithic Fragments	
LSS	Siltstones: Granular aggregates of equant subangular to rounded silt-sized grains with or without interstitial cement. May be well to poorly sorted, with or without sand-sized grains. Composition varies from quartzose to lithic-arkosic, with some mafic-rich varieties.
LSA	Argillaceous: Dark, semiopaque, extremely fine grained without visible foliation, may have mass extinction, variable amounts of silt-sized inclusions, representing shales, slates, and mudstones.
LSCH	Chert: Microcrystalline aggregates of pure silica.
LSCA	Carbonate: Mosaics of very fine calcite crystals with or without interstitial clay- to sand-sized grains. Most appear to be fragments of soil carbonate and are subround to very round.
GROG	Sherd temper: Dark, semiopaque angular to subround grains with discrete margins, including silt to sand size temper grains in a clay, iron oxide, and/or micaceous matrix. The grains differ in color and/or texture from the surrounding matrix of the "host" ceramic. *This parameter is counted only in sherd samples.*
Metamorphic Lithic Fragments	
LMM	Microgranular quartz aggregate: Non-oriented polygonal aggregates of newly-grown strain-free quartz crystallites with sutured, planar, or curved grain boundaries.
LMF	Foliated quartz aggregate: Planar-oriented fabric developed in mostly strained quartz crystals with sutured crystallite boundaries. Quartzite.
LMA	Quartz-feldspar (mica) aggregate: Quartz, feldspars, mica, and opaque oxides in aggregates with highly sutured grain boundaries but no planar-oriented fabric; some are schists or gneisses viewed on edge, some are metasediments or metavolcanics.
LMT	Quartz-feldspar-mica tectonite (schists or gneisses): Quartz, feldspars, micas, and opaque oxides with strong planar oriented fabric. Often display mineral segregation with alternating quartz-felsic and mica ribbons. Grains are often extremely sutured and/or elongated.
LMTp	Phyllite: Like LMT but the grains are silt-sized or smaller, with little or no mineral segregation. Also argillaceous grains which exhibit growth of planar oriented micas silt-sized or smaller.
LMVF	Metamorphosed volcanic rock such as rhyolite. Massive to foliated aggregates of quartz and feldspar grains with relict phenocrysts of feldspar.
LMSS	Metamorphosed sedimentary rock, such as a meta-siltstone. Massive fine-grained aggregates of quartz and feldspar with or without relict sedimentary texture.
LMAMPH	Amphibolite: A high grade metamorphic rock made up largely of amphibole.

Parameter	Description
Unknown and Indeterminate Grains	
UNKN	Grains that cannot be identified, grains that are indeterminate, and grains such as zircon and tourmaline that occur in extremely low percentages.

[1]The coarse foliated rock point count parameters were originally developed to point count coarse muscovite schist tempers (commonly called "micaceous schist") made in the middle Gila River valley. In the Tucson Basin, the Santa Catalina Mountains have schist and phyllite from the Pinal Schist formation, as well as gneiss, which can appear schist-like, from the metamorphosed granitic rocks of the metamorphic core complex. The parameters retain their "S" designation, though the definition is broadened in this table to include rocks other than schist.

CHAPTER 8

SOIL SOURCES FOR CERAMIC PRODUCTION IN THE ANDES

Isabelle C. Druc
Department of Anthropology, University of Wisconsin-Madison .

Andean potters, past and present, have used local materials to produce their ceramics Petrographic and chemical analyses of raw materials and ceramics show the exploitation of a variety of rock and soil sources, up to 12 km away from the place of production. One to three and sometimes four soils can be mixed after grinding and sieving the material to eliminate the coarser fractions. The pots are hand-built and fired on the ground. Soil and paste analyses by petrography and physico-chemical methods allow to explore the variability of the resources, while ethnoarchaeology help assess the geographical extension and use of potting resources Ethnographic examples also show that several villages exploit the same resources and exchange materials. This pattern obliges to reconsider the notion of production area.

The present study was motivated by a need to improve our understanding of local materials and resources available to potters near the ancient ceremonial center of Chavín de Huántar (ca 900-200 BC), in the upper Andean valley of the Mosna river, east of the White Cordillera. The closest potting villages in the region are Yacya, Mallas and Acopalca, above the little town of Huari 44km north of Chavín. West of the White Cordillera is the Santa upper valley, called the Callejón de Huaylas, where much ceramic is produced (see Figure 1). Other potters were also visited near Chacas and San Luis. All these villages are located in the mountainous regions of the Ancash Department.

The potters in several of these villages temper their paste with shashal, a slate-like material. Elsewhere they mix different clayey soils. Manufacture is also different according to the region. In the highlands and on the coast, the materials used for ceramic production are of mineral origin, extracted from local resources. In the tropical forest of Peru, organic material or grog is added to the clay base. Prehistoric and contemporary Andean potters have exploited and still adapt local materials, by modifying their plasticity or granulometry, to produce their ceramics. Whatever the place, local geology and the environment determine what is available. Petrographic and chemical analyses of raw materials and ceramics (ancient and modern) show the use of a variety of soil sources by highland potters. Volcanic fragments, intrusive material, crushed diorite, ground slate, coarse clay, shale-rich soils are examples of the materials used. The types of materials also vary through time, along with cultural and technological changes of production. How was it done in the past? How large was the resource area? What is the regional variability of the raw material used? Soil survey and paste analysis yield some answers to these questions, but ethnographic examples show the need to consider complex production scenarios.

A diachronic perspective of ceramic production in the north central highlands will first be presented. This will allow us to envision the changes observed in soil sources and ceramic pastes in the region of study. We will then focus on ethnoarchaeological data about materials and soil sources for ceramic production around Huari, Ancash. The ethnographic context helps understand the complexity of ceramic studies when several materials are used and the implication of this multiple source use on petrographic and chemical analyses for archaeology. The difficulty of distinguishing potting villages using the same resource areas is then highlighted by the study of the slate material used as temper.

MATERIALS AND RESOURCES

During the first millennium BC, called the Early Horizon (900-200 BC), the ceramics from coastal and Andean sites were usually made from coarse material, alone or mixed with soils with different degrees of plasticity. These soils are derived from sedimentary, intrusive, volcanic or metamorphic sources. Very rarely is pure clay used. The type of paste is in part related to the type of ware produced and ceramic technology. Finer or 'refined' material is used for the production of fine bottles and bowls, but exceptions are known. This pattern is observed in the ceremonial center of Chavín de Huántar. The ceramics from this site display a wide variety of pastes, local and non-local, reflecting the importance and activity of this ceremonial center.

The site is located in the upper Mosna valley, at 3200 m. elevation, east of the White Cordillera. This high mountain range is formed of a granodioritic Batholith, intruded into volcanic and sedimentary sequences (Egeler and de Booy 1956; Cobbing et al. 1981). Alluvial and glacial deposits are found on the slopes of the Cordillera and in the valley bottom, while glaciers cover the peaks above 5000 m elevation. Local geology around Chavín is reflected in the pastes of the ceramics. In the early stages of Chavín, a volcanic paste with welded tuff fragments, green hornblende, biotite, zoned plagioclase and embayed quartz, was used for utilitarian ware and some fine ceramics. Later, intrusive granodioritic material is preferred. Both materials can be found in the White Cordillera, west and south of Chavín. The compositional variations observed in these paste groups suggest the existence of a plurality of workshops, located on the slopes of the White Cordillera and, probably, in the valley floor (Druc 1998).

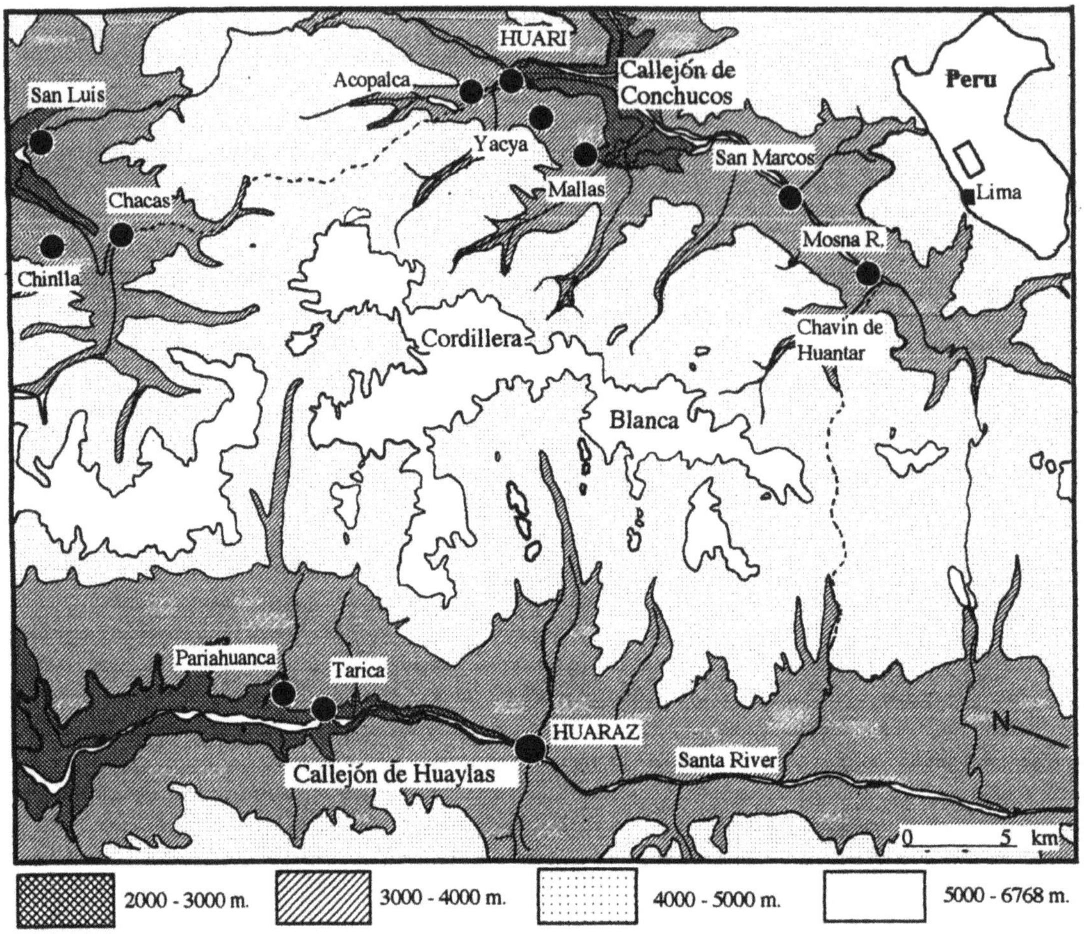

Figure 1. Region of study: Mosna Valley and Callejón de Huaylas.

During the Early Intermediate Period (200 BC - AD 600), in the Callejón de Huaylas and in the Mosna valley, ceramics of Huaras and Recuay (or Huaylas) styles are found post-dating Chavín materials, in particular in Chavín de Huántar (Bennett 1938; Tello 1960; Burger 1984). Late Early-Horizon Huaras and Early Intermediate Recuay cultures are based in the North-Central highlands, in the Santa valley and Callejón de Huaylas (Amat 1979). Their ceramic style and paste composition, however, are much different from that of Early Horizon Chavín. The Huaras style is distinguished by white-on-red painted surfaces, and Recuay wares by negative painting, and red-on-white paint over a fine homogeneous, kaolin paste (Bennett 1944; Tello 1960; Kauffmann Doig 1973). Some Recuay sherds I analyzed, however, have a shale-rich paste and no kaolin clay.

Elsewhere, and particularly on the coast, ceramic production intensifies during this period and the use of the mold becomes frequent. This is seen in the manufacture of figurines at the Moche capital on the north coast, for which local clay beds are mined (Chapdelaine *et al.* 1995). Fine wares and figurines are made of finer material and production is more standardized than for utilitarian wares, which display a wide diversity of paste formulae, particularly in cooking pots. This pattern suggests a multiplicity of production centers or workshops at the domestic level for local consumption, while fewer, centralized centers produce finer ware with a more standardized technology. This is observed for Moche and Cayash ceramics on the coast and in the Cuzco region (Chapdelaine et al. 1995; Krzanowski 1986; Ixer and Lunt 1991).

During the Middle Horizon (AD 600-1000), ceramics with Wari influence (a culture based in the southern highlands) are found in the Callejón de Huaylas and around Chavín. Some of these wares have shale-rich compositions, while others have a detritic temper with granitic and quartzite components, and very few shales (Druc, ms). A gabbro-diorite paste, with pyroxene, amphibole, quartz and plagioclase is also observed in other post-Chavín (#3760a,b,c) and Middle Horizon wares (#C22, C24). This intrusive rock paste-type is reminiscent of the granodioritic type found in several Chavín ceramics, suggesting a continuity in the use of the resources along the White Cordillera. The other types of pastes produced during Chavín times, however, are not found, and ceramic production must have been very low in Chavín in later periods, or nonexistent.

During the Late Intermediate Period (AD 1000-1476), Inca Domination (Late Horizon - AD 1476-1534), and the Colonial Period, no fine clay appears to have been used in the North-Central highlands. Colonial influence is seen in the appearance of new ceramic forms, but not in technology of manufacture. It is not yet known when the use of slate as temper was introduced (see Druc 2001 for a discussion on this matter). It is used in modern production around Huari and Chinlla, but occasional slate fragments are seen in earlier wares.

Finally, modern potters from Taricá and Pariahuanca in the Callejón de Huaylas mix two to four soils of different plasticity, after grinding and sieving them to eliminate the coarser fractions (Druc 1996). Typically, the potters use yellow and black earths, the former having more silt and quartz grains, the later more shales (Druc and Gwyn, 1998). They mine the black earth from deposits at the limit of ancient glacial moraines, at 3200 m altitude. In traditional production, pots are hand-built with coils, walls are straightened with the paddle-and-anvil technique and the wares are ground fired. Only cooking pots (ollas), jars, and maize toasters (*cancheros*) are made. Bowls and bottles are not produced anymore. Ceramic technology is different in the Huari region. There, the potters use a yellow plastic earth and ground slate, which they extract from old mines. The coiling technique is used, but thicker coils are made than in the Callejón de Huaylas, and the paddle-and-anvil technique is not employed. The resources in both areas are often 6 to 12 km away from the production site.

This brief overview of ceramic production through time in the North-Central highlands highlights the changes in paste composition and resources used. Figure 2 summarizes these changes for the Chavín area and the Callejón de Huaylas.

On a more local basis, the ceramic styles found throughout the history of the Mosna valley reflect the influence of different cultures. We see a move from a plurality of paste types and resources at Chavín de Huántar during the Early Horizon to a simpler, less diversified production in later periods. This probably reflects the shift in importance of the site. The change includes the use of different soil and temper resources; in particular the use of crushed black shale or carbonaceous material as temper. It is referred to as graphite temper or slate temper (*piedra pizarra*) according to the region. However, the terminology may refer to different materials. The study of actual resources and production strategies allows us to propose scenarios for interpreting the archaeological data. This may not be applicable to the distant past when the Chavín culture was flourishing, but analysis of later ceramics shows the use of materials, which are still mined today. The next section deals with this problem and presents some ethnographic data for the Huari region.

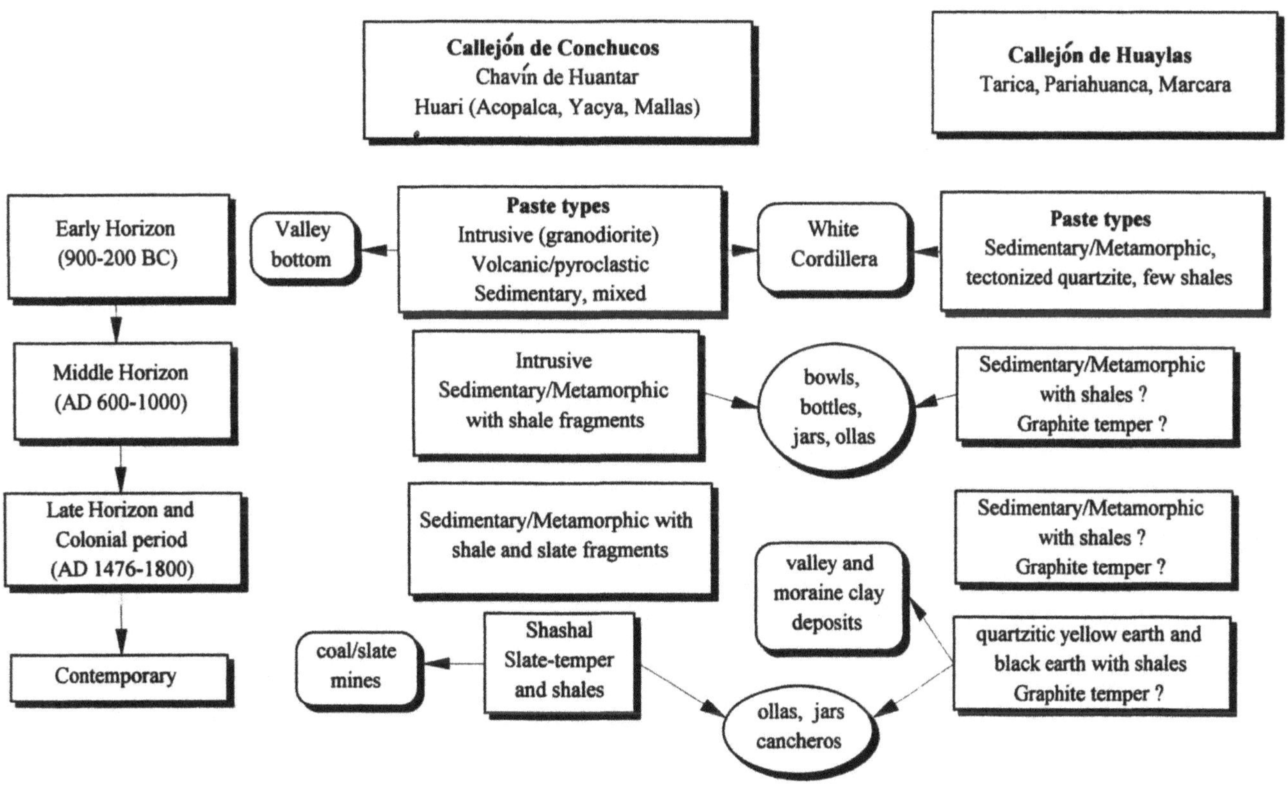

Figure 2. Diachronic perspective of soil sources and ceramic production in the north-central Andes.

GRAPHITE-TEMPERED WARE

Several late pre-Hispanic and modern wares are termed graphite-tempered or slate-tempered, in reference to the black carbonaceous material, graphite or slate fragments heavily tempering their paste. Local people sometimes call them stonewares (*ollas de piedra*). They are known to be more shock-resistant and heavier than the ceramics from Taricá, in the Callejón de Huaylas. Graphite temper is reported in utilitarian jars from the modern levels of Piruro (Pan 5-8) a Pre-Hispanic site in the Callejón de Huaylas and in Taricá pots used by modern shepherds (cf. Ponte Rosalino, personal communication 9/1998). Victor Ponte Rosalino also mentions the existence of graphite-tempered ceramics from Huanuco Pampa, a provincial Inka administrative center south of La Union, above the Marañon River. John Rick found similar temper in ceramics from the upper levels of the temple of Chavín de Huántar (personal communication 9/2000). However, the terminology regarding this type of temper is unclear. No graphite or slate has been observed in the samples I analyzed from Taricá or the late ceramics from Chavín de Huántar. Lupe Camino (1984), who conducted ethnographic work in Taricá, does not mention either the use of slate or graphite temper. Hence, the terms graphite or slate-temper may refer to different types of crushed materials. The analysis and exchange of thin sections would clarify the terminology.

SLATE-TEMPERED WARE

Only few examples of slate-tempered ware are found in the literature. The use of slate temper was observed in Yacya, Mallas, Acopalca and Chinlla (Druc 1996, ms). In the regions of Huari and San Luis-Chacas, the slate-temper is called *shashal* and the yellow clay-base *raku*. These terms are widely used and may refer to different materials, hence the terminology problem found in the literature. Ixer and Lunt (1991) report the use of slate temper in the southern Cuzco area. The slate is described as 'lath-shaped, limonitically-stained fine-grained metapelite (metamorphosed shale) up to 3 mm in length' (ib: 148). Ravines (1989) reports the use of slate-temper (*piedra pizarra*) in Cusca and crushed rock-temper in Aco (both in the Corongo Province, Ancash Department). Crushed rock is also used in Caulimalca (Otuzco Province, Department of La Libertad), in Cachipampa and Marcabal Grande (*piedra shalar*) in the Province of Sánchez Carrión (La Libertad). However, Ravines does not give a description of the material and it is difficult to know the real composition of the temper.

COMPARATIVE ETHNOARCHAEOLOGICAL DATA

To characterize the modern resources, and in particular the slate-tempered wares, an ethnoarchaeology study was conducted. Ethnoarchaeology combines ethnographic study with material analysis to provide archaeologists with comparative data and scenarios to interpret the archaeological data. Mineral and chemical analyses of modern pastes allow us to verify the extent of soil and paste variability in the region of study and the local potential for ceramic production. However, potters' behavior regarding raw material acquisition may alter the success of material analysis (Velde and Druc 1999). This is observed with the potters around Huari, which exploit resources outside their community and exchange resources with other potting villages.

Figure 3. Crushed shashal (right, next to the wall) and grinding stone (back), Yacya, Ancash.

The three production villages of Yacya, Mallas and Acopalca are located within an 8 km radius around and above the town of Huari, between 3300 m. and 3500 m. of elevation. However, the measure of kilometers does not provide a proper appreciation of travel distances in these mountainous areas. It takes 20 to 30 minutes by foot to reach Acopalca from Huari, one hour and 15 minutes to hike to Yacya and about two hours to get to Mallas. Mallas can also be reached from a trail branching off from the San Marcos-Huari road along the Mosna River. A few potters produce once or twice a year, but this was more intense in the past.

The shashal mines used for obtaining temper are 6 to 8 hours away by foot above Acopalca, and pertain to this community. The shashal is crushed with a large grinding stone (Figure 3) and mixed to the clay or raku. The raku material is found near Yacya. Thus, the potters from Yacya get their shashal above Acopalca, and the Acopalca potters get their raku just below Yacya. There is a tacit exchange of material between the two communities. The Mallas potters use local raku and shashal from mines in the highlands above the village. Occasionally, potters from Yacya also get shashal from local mines above the village. The acquisition pattern observed around Huari is illustrated in Figure 4.

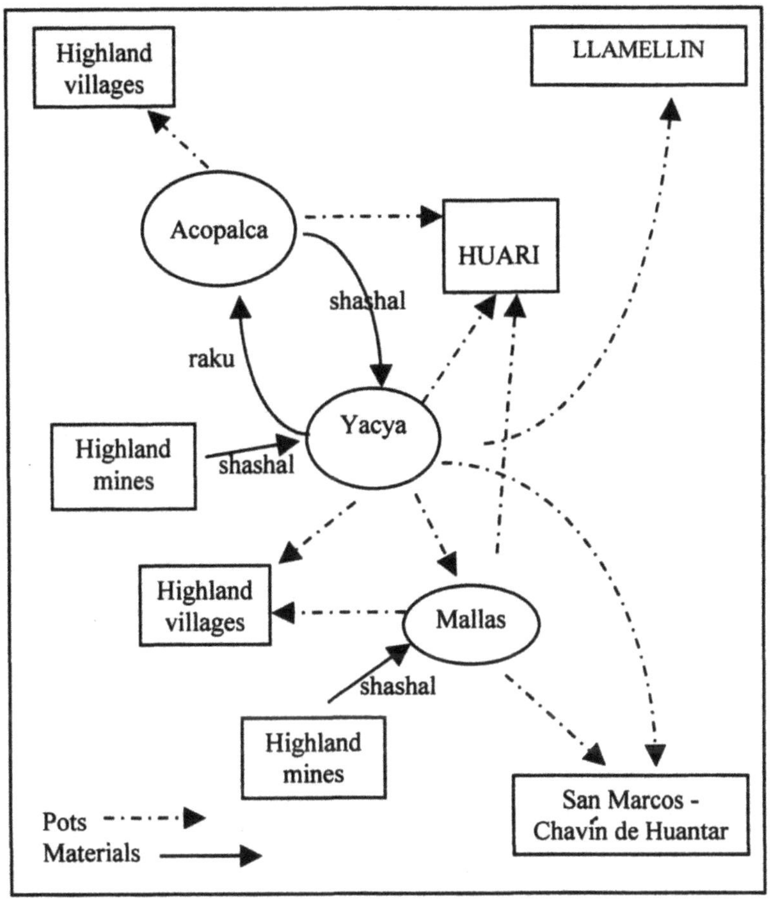

Figure 4. Material acquisition pattern and ceramic distribution around Huari, Ancash.

In terms of source variation, as seen from petrographic and carbon analysis, the shashal mines above Acopalca have similar compositions (the slate is very similar from one mine to the other), while the raku composition shows more variability. The raku is an argillaceous soil extracted from different fields of the community of Yacya, with a variable distribution and granulometry of quartz, feldspar, quartzite, clastic and terrigenous constituents.

The pattern illustrated above makes it very difficult to distinguish the ceramic production based on composition between the producing villages within a 15km perimeter (approximately a 25 square km area). Furthermore, forms and ware decoration tend to be similar, with individual decorative variants from one potter to the other (Figure 5). The petrographic analysis reported in the next section shows how the extent of the production zone as seen from mineralogy differs from reality.

Figure 5. Types of pots produced in the Huari region.

In terms of ware provenance, the ethnographic pattern of ceramic production observed around Huari suggests the following attributions (Figure 6):

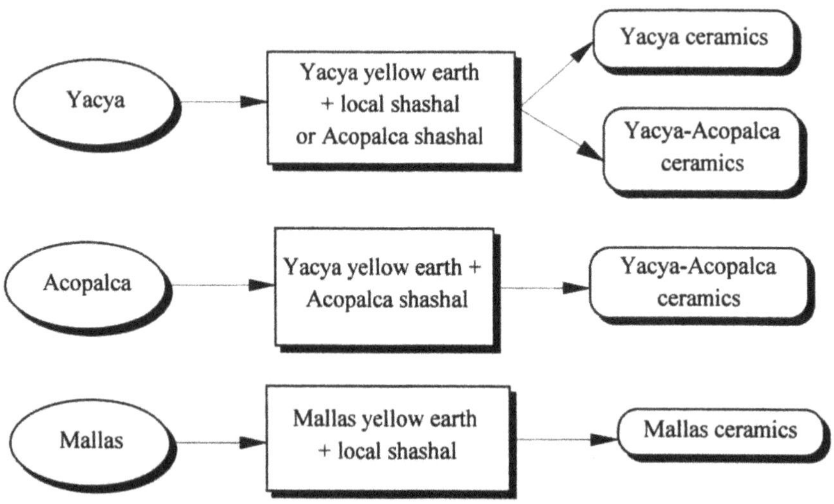

Figure 6. Ethnographic pattern of paste preparation and theoretical provenance attribution.

In archaeological context, however, the potters' behaviors are unknown and we rely only on material remains and analysis to deduce production scenarios. In the situation described above, the theory based on paste analysis would predict three distinct provenances. In practice, petrographic and chemical analyses only poorly distinguish the different villages involved. Thus, a large production area must be considered, which includes the producing villages sharing the same resources. This production zone reaches 25 sq. km. A similar situation occurs in the Callejón de Huaylas, for the villages of Taricá, Pariahuanca and Marcara.

In order to see if mineral and chemical differences are observed between the materials and products of the ceramic villages around Huari, several analyses were conducted: petrography, carbon analysis and chemical study by SEM and XRF. These analyses are described in the following sections.

PETROGRAPHIC ANALYSIS OF MODERN POTTERY PRODUCTION

The petrographic corpus consists of 31 thin sections of unfired raw material and ceramic fragments collected in 1997 and 1998, from different potter communities around Huari, San Luis-Chacas, Chavín and Huaraz in the north-central Andes. In addition, 24 samples were taken from post-Chavín ceramics (Recuay, Middle Horizon, colonial period) from Chavín de Huántar and Yacya for comparative purposes and verification of continuity in paste formulae with the existing data set from Early Horizon Chavín ceramics (Druc 1998).

The analysis of the non-plastics distinguishes four groups, related to the four regions sampled: Huari, Chacas-San Luis, and Huaraz. This classification shows that material composition differs from one region to the other, although shashal is used in the first two regions and shale-rich soils in Chavín and Huaraz. However, the analysis does not precisely discriminate the production villages within the same region. We present here the results for Huari and San-Luis-Chacas being concerned with distinguishing centers that use similar materials (slate temper). The analysis results for Chavín and the Callejón de Huaylas have been presented elsewhere (Druc 1998; Druc and Gwyn 1998).

In the Huari region, ceramics from Mallas, Yacya and Acopalca constitute two intermixed groups, with slightly dissimilar mineral compositions (Plate 2, figures e and f). Group 1 is characterized by metamorphic rock fragments with fine white mica plates and fragments of shale and slate. Group 2 only presents the shale and slate fragments. Group 1 (S4, S7, and S10) combines old and used olla fragments from Acopalca and Mallas, and one unfired paste mix from Yacya. Group 2 (S1, S6, and S11) is comprised of ollas from Acopalca, Mallas, and one old olla fragment from Yacya. Acopalca S1 and Mallas S6 are very similar in texture, siltosity and granulometry. They have a fairly unsilty and gray matrix. Yacya S11 can be singled out for its silty matrix, the presence of quartzite fragments and many quartz inclusions, and oxides and black filaments covering the rock fragments. Two raku samples (S3 and S9) could not be classified. Both come from clay deposits near Yacya, but raku S3 was given by a potter from Acopalca and S9 by a potter from Yacya. Both soils show mineral similarities, but not strongly enough to be included in one of the two groups. S3 is a soil of very fine granulometry with mica but with no lithics. S9 could be included in Group 1 because of the presence of two metamorphic fragments with mica, but the slate and shale fragments characteristic of Group 1 are missing.

As observed elsewhere (Druc and Gwyn 1998) this classification demonstrates that ceramics best compare to each other, while unmixed and unfired raw materials are more difficult to relate to the finished products. Furthermore, the classification based on non-plastics does not clearly distinguish the three ceramic centers around Huari. The exchange of raw materials explains the constitution of groups with samples of different site provenance. The slate is present in both groups, while the raku serves to differentiate the two groups. This differentiation is grounded in the compositional variability of the raku material in the community of Yacya. The presence of Acopalca sherds in the two groups suggests that the potters have used two different places of extraction (be it two potters, each using a different source, or one potter acquiring his material at different places over a certain time between two productions). The same is observed for Yacya and Mallas.

Only one group was constituted from the ceramics from the region of Chacas-San Luis (Group 3). In this petrogroup, the slate fragments are less opaque, of lower metamorphism than the slate fragments from the Huari region. They have the appearance of shale. Intragroup variability is very small. All thin sections from Chinlla and Ichikchinlla, collected from two different potters, are very similar in composition, texture, granulometry and color, except for S13, which has a darker paste.

Given that all the wares in these two regions are slate-tempered, how do they classify according to the chemical characterization of the slate? Carbon analysis of the slate fragments from different mines was conducted to seek information on the regional variation of slate composition. SEM analysis of slate fragments in ceramics was also conducted, as well as bulk analysis of powder samples of slate-tempered ceramics.

CARBON ANALYSIS OF SLATE SAMPLES

Six samples of raw slate fragments from Yacya, Acopalca, and Chinlla near San Luis-Chacas, were submitted to carbon analysis to test regional tendencies in carbon content. Chang Soobun of the Geology Department at Yale ran the analysis on a carbon detector LECO CR12. Steven Petsch and Isabelle Druc reduced the samples to powder using a mechanical grinder. Carbon determination was based upon burning of the sample, which produced CO_2, to be detected by infrared spectrometry. The calibration was done in reference to standards of calcium carbonate powder $CaCO_3$ 12.00 wgt % from Mallkinckrodt Chemical Works in Saint Louis. The same standard was measured three times, at the beginning, middle and end of the analysis to control for drift and to measure the analytical error. In the present case, the error ranged between 0.7 to 0.8 %.

The six samples were run two times, ca 100-150 mg powder at a time. The powder was poured in an alumina container and inserted into the burning chamber for 2 to 3 minutes or until combustion of carbon present in the sample was complete. The accumulated CO_2 was measured, divided by sample weight, and multiplied by an internal coefficient determined by the calibration program. The analysis yielded the amount of total carbon measured. The total carbon measured per sample (two runs each) is presented in Table 1.

Table 1. Total carbon measured in raw slate samples from Huari-San Luis region

sample	provenance	weight	total carbon	Mean	C (real)	comment
slate 2	Acopalca,	160 mg	57.62 %	56.49 %	55.52 %	slow burning
slate 2/2	Cro Matadero	133 mg	55.35 %			
slate 8	Yacya,	117 mg	53.69 %	53.69 %	52.76 %	Id.
slate 8/2	Cro Ventanilla	123 mg	53.69 %			
slate 66	Yacya,	113 mg	76.32 %	76.54 %	75.22 %	Id. coal?
slate 66/2	mine	145 mg	76.75 %			
slate 14	Ichicchinlla,	120 mg	1.63 %	1.56 %	1.53 %	fast burning,
slate 14/2	Cro Chinlla	123 mg	1.48 %			black shale?
slate 17	Chinlla,	129 mg	1.62 %	1.81 %	1.77 %	Id.
slate 17/2	Cro Chinlla	121 mg	1.99 %			
slate 18	Chinlla,	119 mg	1.001 %	1.09 %	1.07 %	Id.
slate 18/2	grounded	176 mg	1.18 %			
std	standard	319 mg	12.24 %			beginning
	CaCO3	332 mg	12.09 %			of analysis
		325 mg	12.20 %	12.21 %		
		268 mg	12.26 %			end
		307 mg	12.24 %			

The real content of carbon is calculated as follows: (1) C % (real) = C % (measured) x standard fixed/standard measured
The fixed content of carbon in the standard is 12.0 %. The mean standard measured is 12.21 %.

Carbon can be present in the sample as different compounds or components such as graphite, calcium carbonate, organic or inorganic (de-gazed). To distinguish if it is organic carbon or graphite that is present in the sample, thermogravimetric methods are useful, as graphite is more robust in crystallinity than carbonate. If two separate peaks appear, this indicates the presence of two different types of material, such as organic carbon and carbonate carbon. The compounds are all organic if the two peaks immediately follow each other (Soobun, personal communication, 4/1998). Graphite is composed of 100 % carbon, anthracite (coal) 70- 90 %, ignite, ca 60 %, while black shale is very low in carbon.

The amount of carbon identifies the materials as coal for slate sample #66 from a mine above Yacya, ignite for samples #2 from Yacya and #8 from Acopalca, both extracted from mines above Acopalca, and black shale for samples #14, #17, and #18, from the Chacas-San Luis region (Figure 8).

All these materials are called shashal, or slate temper, but a more exact denomination would be coal temper and shale temper. The graphite-tempered ceramics from Taricá and Chavín, reported by Ponte Rosalina and John Rick, could thus respond to the same denomination, while being different in composition from the shashal from Huari. The amount of carbon also differs according to the resource area and, probably, the vein. The latter should, however, be verified with more samples to assess the internal variability of each source. Nevertheless, we can conclude that striking differences exist between the shashal temper used in the Huari region by the potters of Acopalca and Yacya and the shashal from San-Luis Chacas used by the potters of Chinlla.

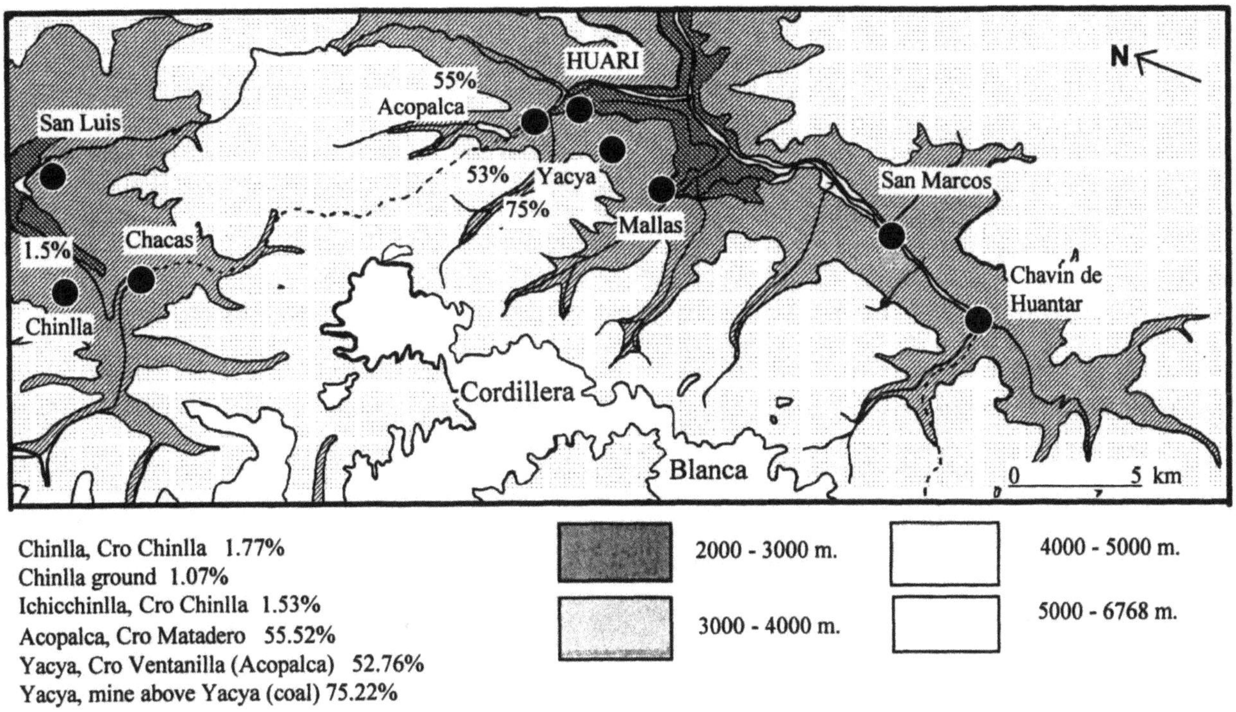

Figure 8. Carbon content in shashal temper in the region of study

EM AND XRF ANALYSES

.s the shashal temper used by the potters differs in omposition from one region to the other, it was thought that nergy dispersive X rays analysis (EDX) using a scanning lectron microscope (SEM) would be a good tool in istinguishing the different production villages. The analysis vas conducted with the help of Pierre Magny at the Jniversité de Sherbrooke, Canada. SEM-EDX allows us to haracterize specific grains instead of the whole paste as in ζ-ray fluorescence or neutron activation. The shashal ragments in the ceramic paste are easily identified by etrography or simply on macroscopic inspection as the ragments are usually very large (2 mm or more). It is thus asy to position the electron beam on shashal fragments to haracterize their composition. As carbon is a light element, pecial parameters are set to measure it, using the ultra thin vindow, 200 analysis seconds, and a probe current of $3\text{-}10^{-9}$, 5 kV and a vacuum of $2\text{-}10^{-5}$mbar. Magnification typically uns between 1400 and 1700.

Γhe analysis used the program Ursa (ultra rapid spectrum analyzer) for calibration. Four ceramic thin sections coated vith Au-Pd (gold-palladium) and two raw shashal fragments hat had already been analyzed by carbon analysis were ubmitted to the SEM-EDX analysis. Only the major elements showing as peaks in the EDX spectrum have been neasured (C, Mg, Al, Si, K, Ca, Ti, Fe). However, the SEM analysis was inconclusive and did not allow discriminate adequately the production of the neighboring villages. The composition of the temper fragments varies from one ceramic to the other and, sometimes, within the same ceramic. Also, carbon content differs from that observed in carbon analysis, The calculation procedure may account for the difference, but other variables should be considered, such as the changes occurring during the firing of the ceramics, be it a loss (in an oxidizing atmosphere) or gain (in a reducing atmosphere) of carbon.

The X-ray fluorescence analysis (EDXRF) of raw shashal fragments and ceramic samples measured 23 major, minor and trace elements. The results of the raw shashal fragments showed regional differences in composition, in particular for the silicon and aluminum contents. As expected from carbon analysis, the composition of the shashal from Chinlla presents a much higher content in SiO_2 and Al_2O_3 (63.83%-26.66%) than for samples from Huari. But the analysis is not satisfactory on a more local scale. The analysis of the powder samples of slate-tempered ceramics show the regional variations, but again, the exchange of materials between production centers hinders finer distinctions. In this regard, neutron activation analysis might yield better results in distinguishing the villages, as it is more sensitive to small compositional variability (minor or trace elements).

CONCLUSIONS

The results of the analyses conducted in this study show how important it is to combine different data sets, for none of the analyses are entirely satisfactory when used alone. On the basis of ethnographic investigation, several ceramic groups and production villages were expected to be identified, with different paste formulae used in relation to the actual pattern of clay and temper acquisition and exchange. This pattern was not identified in ceramic analysis, resulting in a production area extending to 25 square kilometers around Huari and including all the villages sharing the same resources. A similar scenario of material exchange between potting villages can be expected for past productions and should be considered when interpreting archaeological data. If the pattern of ceramic production observed around Huari and in the Callejón de Huaylas is applied to the upper Mosna, a 25 sq. production area around Chavín can be proposed, with a resource area 6-12 km away from the producing villages. This yields a production zone that includes San Marcos, Huántar, and the Mosna drainage, up to the sources of volcanic materials reported in the geological maps. Several hamlets and small villages around Chavín and in the Mosna drainage have been reported by earlier surveys (Tello 1960; Amat 1979). Some of these villages could have played a role in the production of local ceramics. Only archaeological excavations of these small settlements could bring some more information on that matter.

In terms of ceramic and material analysis, petrography identified petrogroups on a regional basis, but not on a village basis within the region. This is explained by the use of the same resources by different, neighboring villages. The ceramics form better groups while raw materials are more difficult to relate to the end product. Carbon analysis of raw slate used as temper discriminated well between the mines and allowed the characterization of the type of material used as coal, ignite or black shale. However, one must remember that in the archaeological context, raw materials are rarely found and ceramic fragments are usually the objects of analysis. In this line of thinking, a SEM-EDX analysis was performed on ceramic thin sections tempered with slate.

The results show differences between the temper composition of each thin section, and also within the same ceramic (for one ceramic from Mallas and one from Yacya). This internal variability can be interpreted as an indication of a plurality of temper sources unless the variability of the raw material source is known, which is rarely the case in the archaeological setting. Hence the need for a survey of raw materials and soil analysis, combining geological expertise and archaeological concerns. In the case above, the temper difference indicated the use of two different slate sources for the preparation of the paste. This is possible with Yacya and Mallas potters, who get their temper material from Acopalca and local mines. However, the mixing of tempers from different sources was not observed during the ethnographic study, nor reported by the potters, although they may have failed to report it. Finally, the X-ray fluorescence analysis of raw slate fragments did show regional differences, but the analysis of powdered ceramics was not as discriminate. Again, the exchange of materials between production centers hinders finer distinctions. More sensitive analytical methods might be more successful.

This situation of material exchange between neighboring potting villages offers a scenario for ceramic production somewhat more complex than in other models. The production area is consequently enlarged to include all the villages sharing the same resources. As demonstrated here and as Dean Arnold already stressed in 1972 in his study of modern ceramic materials near Ayacucho, the interpretation of petrography and chemical analysis is greatly helped by referring to ethnographic scenarios and the evaluation of resource variability within a region. For this matter, material analysis and collaborative work with clay mineralogists and geologists, or a good knowledge of the local geology are advised.

Acknowledgments

This research was made possible by a postdoctoral grant from the Social Sciences and Human Resources Council of Canada. My acknowledgment also goes to Chang Soobun and Steven Petsch of the Geology Department at Yale for the carbon analysis, and the museums and institutions that provided the ceramic samples and the logistical support to conduct the analysis. In particular I want to thank Wilder Leon Ascurra in Peru, Richard Burger at Yale University, Pierre Magny of the Service de microscopie électronique, Faculté de médecine, Université de Sherbrooke and Regina Zamojska, Département de Chimie, Université de Sherbrooke.

References

Amat, H. 1979. Arqueología Ancashina. *Cuadernos de difusion* 17: 27:39, Instituto Nacional de Cultura, filial Ancash.

Arnold, D. E. 1972. Mineralogical analyses of ceramic materials from Quinua, Department of Ayacucho, Peru. *Archaeometry* 14, 1: 93-102.

Bennett, W. C. 1938. *1938 Expedition, Callejon de Huaylas, Department of Ancash, Peru.* Unpublished manuscript, American Museum of Natural History, New York.

Bennett, W. C. 1944. *The northern highlands of Peru: Excavations in the Callejon de Huaylas and at Chavín de Huántar.* Anthropological Papers of the American Museum of Natural History 39(1). New York: American Museum of Natural History.

Burger, R. L. 1984. *The prehistoric occupation of Chavín de Huántar, Peru.* Anthropology 14. Berkeley: University of California Press.

Camino, L. 1984. Tarika, un centro alfarero. *Boletin de Lima* 35: 49-53.

Chapdelaine C., Uceda S., and Kennedy G. 1995. Activación neutronica en el estudio de la producción local de la ceramica ritual en el sitio Moche, Peru. *Bulletin de l'Institut français d'études andines*, 24: 183-212.

Cobbing, J., Pitcher, W.S., Wilson, J.J., Baldock J.W., Taylor, W.P., McCourt W., Snelling, N.J. 1981. *The geology of the Western Cordillera of Northern Peru.* Overseas Memoir 5. London: Institute of Geological Sciences.

Druc, I. C. 1996. De la etnografía hacia la arqueología: Aportes de entrevistas con ceramistas de Ancash (Peru) para la caracterización de la cerámica prehispánica. *Bulletin de l'Institut français d'études andines* 25(1): 17-41.

Druc, I. C. 1998. *Ceramic production and distribution in the Chavín sphere of influence.* Oxford: British Archaeological Reports, international series 731.

Druc, I. C. 2001. Shashal o no shashal esa es la cuestión. Etnoarqueología cerámica en la zona de Huari, Ancash. *Bulletin de l'Institut français d'études andines.*

Druc, I. C., and Gwyn, H. 1998. From clay to pots. *Journal of Archaeological Science,* 25(7): 707-718.

Egeler, C.G., and De Booy T. 1956. Geology and petrology of part of the southern Cordillera Blanca, Peru. *Geology Serie* 19(1), 1-86.

Ixer, R. A., and Lunt, S. 1991. The petrography of certain pre-Spanish pottery from Peru. In *Recent developments in ceramic petrology,* ed. A. Middleton and I. Freestone. British Museum Occasional Papers, no 81, pp. 137-164.

Kauffmann Doig F. 1973. *Manual de arqueologia peruana.* 5th edition. Lima: Peisa.

Krzanowski, A. (Ed.) 1986. *Cayash Prehispanico.* Polska Akademia Nauk, Prace Komisji Archeologicznej, no 25.

Ravines, R. 1989. Principales comunidades y centros alfareros del Peru. In *La ceramica tradicional del Peru,* ed. R. Ravines and F. Villiger. Lima: Los Pinos, pp. 46-57.

Tello, J. 1960. Chavín cultura matriz de la civilizacion andina. Lima: Universidad Nacional Mayor de San Marcos.

Velde, B., and Druc, I. C. 1999. *Archaeological ceramic materials. Origin and utilization.* Berlin Heidelberg: Springer Verlag.

CHAPTER 9

DEVELOPING TECHNOLOGICAL STYLES OF PETÉN POSTCLASSIC SLIPPED POTTERY

Leslie Cecil
Department of Anthropology, Southern Illinois University

Different technological styles of Petén Postclassic slipped pottery were developed from mineralogical (petrographic thin-section and x-ray diffraction) and chemical (strong acid-extraction ICPS) analyses of clay minerals, as well as stylistic analysis. These styles define the variability of technological characteristics in pottery manufacture. The variability in the mineralogical and chemical signatures of Petén Postclassic slipped pottery reflects the combination of technological activities or behavioral events that underlie social identity. These manufacturing choices support the proposition that Postclassic Petén potters (the Itzaj and Kowoj) produced and reproduced pottery technological styles as part of their social identities in a socio-political milieu of changing alliances, changing dominance relations, and repeated migrations.

On the eve of the Spanish conquest, there were two dominant and mutually antagonistic social groups in the Petén lakes region—the Itzá and the Kowoj. Each social group had a principal province, headed by a leader at a province capital, controlled different subprovinces, and had origin and migration myths from which they claimed ancestry (Figure 1).

In this context, it is important to make use of the Postclassic (A. D. 950-1697) period pottery of the lakes area to delineate differences among social identities, especially between sites/territories of the Itzá and the Kowoj, because antagonistic social/ethnic groups may use «style» to distinguish themselves from «the other» (Hodder 1977; Sackett 1985; Wiessner 1983, among others). One way to describe differences in social identities is through the analysis of the technological styles of Petén Postclassic slipped pottery. Technological style, as defined by Lechtman (1977; see below), provides an approach to examine the style of material culture that describes the relationship between design elements, the patterns people produce through various behaviors, and the extent to which social/ethnic identity is reflected in the «style» of that object. When creating material culture, such as pottery, the producer may have many choices (operational sequences) to make that reflect sociological and cultural constructs that underlie and direct her/his actions, and subtle differences in a choice can influence the social representation of material culture. Therefore, both the material and the process of manufacture contribute to an object's style as much as does the surface decoration because technological acts are embedded in symbolic systems that reflect social reality, indigenous knowledge, and social/ethnic identity.

The technological acts reflect social/ethnic identity because Maya potters are social actors in a social structure that embodies their identity. Choices made during the manufacture of pottery (e.g., clays, mineral inclusions, vessel form, decoration, etc.) reflect the patterned actions or tendencies of manufacturing techniques (Bordieu 1977). This type of patterned activity embedded in structure and agency allows the archaeologist to study materials such as clay and mineral inclusions and patterns of pottery manufacture such as vessel form and decoration as practice. As such, the patterns of manufacture are not "'added on' in order to signal group identity," but are "the process by which a sense of group identity is formed and transformed as being coeval with and identical to the process by which a sense of techniques is formed and transformed" (Dietler and Herbich 1998: 247).

Although the entire patterned activity of pottery manufacture is important, sherd paste data are emphasized because of the focus of this volume. This type of analysis demonstrates that archaeologists can use pottery clay paste data to go beyond clay mineralogical and archaeometric studies in order to answer broader archaeological and anthropological issues such as social identity. Decorative characteristics are presented briefly in order to establish technological styles that reflect social identity. As a result of mineralogical and stylistic analyses, the technological styles of Petén Postclassic pottery demonstrate that: 1) technological and stylistic choices have a social context; 2) technology and style are social reproductions of Postclassic society; 3) some technological and stylistic choices were more compatible than others within Postclassic Maya society; 4) technology affects style; and 5) these compatible choices reinforced the existing technology and social ideology. Ultimately, similarities of formal arrangements of technological and stylistic patterns associated with different ceramic groups and/or archaeological sites refines our understanding of the settlement and sociopolitical relations of the Itzá and the Kowoj.

ETHNOHISTORY OF THE ITZÁ AND THE KOWOJ

At the time of the Spanish conquest, Petén was divided into administrative provinces that were headed by and named after a dominant lineage and each lineage had control of several subprovinces or territorial regions throughout the area (Jones 1998). Two prominent, warring ethnic groups know by their dominant or ruling lineage name are said to have occupied the Petén lakes region at the time of the Spanish conquest: the Itzá and the Kowoj (Jones 1998: 18-19).

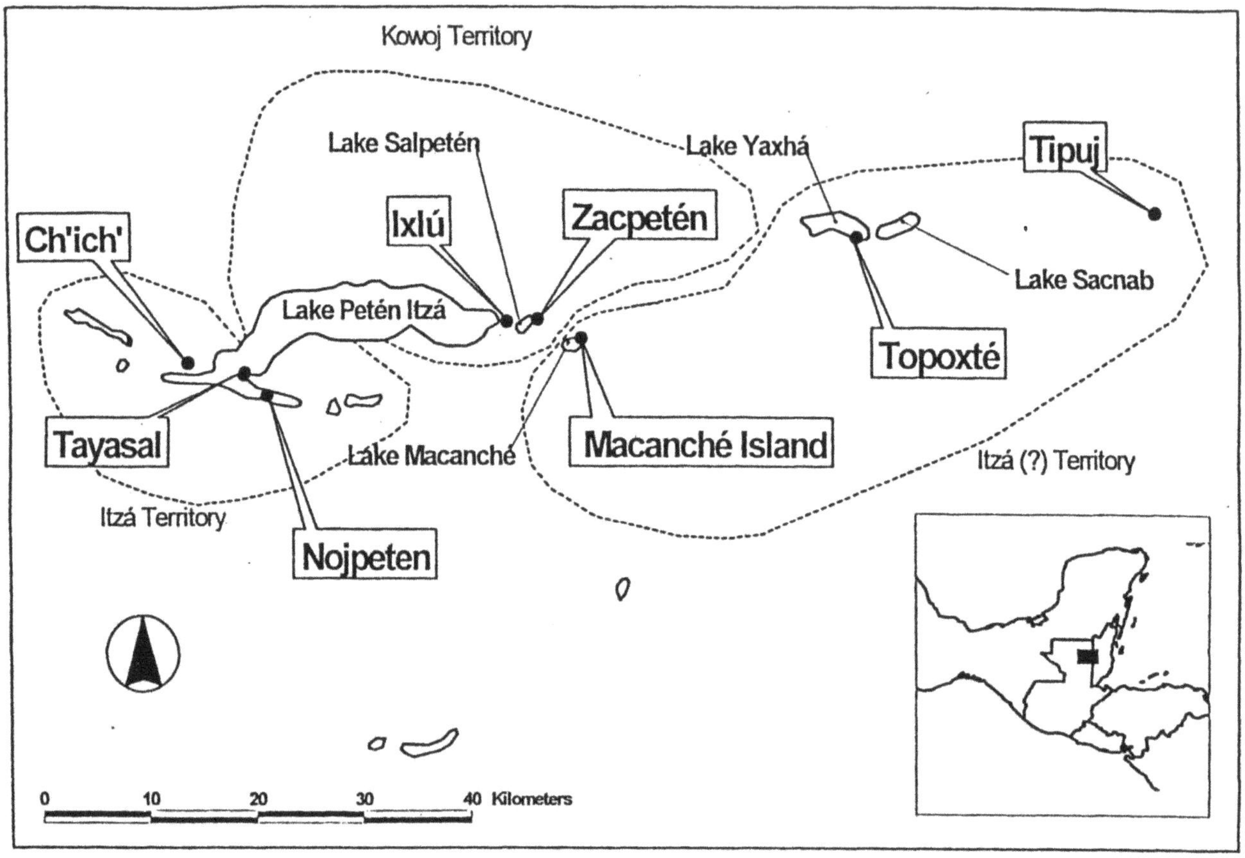

Figure 1. Postclassic Archaeological Sites and Ethnohistorical Social/Ethnic Groups (adapted from Jones 1998: Map 3)

Regardless of their migration history, the Itzá were present at Lake Petén Itzá when Cortés traveled through Petén on his way to Honduras (Cortés 1976: 219-285). At this time, the Itzá capital was Nojpeten or Taj Itzaj, which was located on present day Flores Island. Spaniards described Nojpeten as having multiple temples and hundreds of houses with the surrounding islands and mainland supporting a large population (Thompson 1951: 390-394). The Itzá dominated territory west and south of Lake Petén Itzá (Jones 1998).

The Kowoj occupied the northern and eastern part of Lake Petén Itzá and Lake Salpetén and claimed to have migrated from Mayapán around AD 1530 (Jones 1998: 18). Although Jones (1998: 17-18) believes that the migration of the Kowoj from Mayapán was a result of Spanish contact and political turmoil around AD 1530, it is possible that they had a series of migrations to and from Mayapán, of which one occurred after the fall of Mayapán. The Kowoj relation to Mayapán is strengthened with the linkage of Kowoj kinship patronmyns and matronyms to prestigious individuals at Mayapán. For example, Roys (1967: 79) notes that in the *Chilam Balam of Chumayel* the name Kowoj was associated with an individual listed as the «guardian of the east gate» at Mayapán. Jones (1998: Table 1) also demonstrates that Kowoj patronyms and matronyms occur in the Petén lakes region during the 17th century.

THE THEORY OF TECHNOLOGICAL STYLE

Although many theories exist to guide the analysis of style of material culture, the theory of technological style (Lechtman 1977, 1979, 1981; Lemonnier 1992, 1993) systematically examines technological/ behavioral attributes along with surface stylistic embellishments to differentiate social/ethnic identity. «Technology» is defined here as the sequence of manufacturing choices that involve matter, energy, tools, and specific knowledge (Lemonnier 1992: 5-6). «Style» refers to the «visual representations, specific to particular contexts of time and place, that...transmit information about the identity of the society that produced the style and about the situation or location where it happens» (Rice 1987: 244).

When creating material objects, the producer may have many choices (operational sequences) to make that reflect sociological and cultural constructs that underlie and direct his/her actions, and subtle differences in a choice (e.g. matter, energy, specific knowledge, etc.) can influence the social representation of material culture (Lemonnier 1993: 9; Lechtman 1977: 6). The material and the process of manufacture contribute to an object's style as much as does the surface decoration because technological acts are embedded in a symbolic system that reflects social reality and indigenous knowledge. Therefore, style is a technological performance that allows the archaeologist to better understand emic behavior based upon primarily etic phenomena of nature (Arnold 1971; Lechtman 1977: 7).

Different technological styles may be developed and operate synchronically, but they will not be perpetuated unless the technological style is compatible «with the natural environment and with the state of technological systems at the time of creation» (Lemonnier 1993: 12). Selective pressures at the individual and community levels decide what will represent social structures such as power, ancestry, and identity (Lemonnier 1993: 15). The integration of behavioral events, choices, and «selective pressures» at each step of manufacture defines technological style and is «recognizable by virtue of its repetition which allows us to see the underlying similarities in the formal arrangements of the patterns of [manufacturing] events» (Lechtman 1977: 7). These persistent arrangements or patterns represent social/ethnic identity.

In order to determine different social/ ethnic groups and the pottery that they produced in the central Petén lakes region, the author examined the similarities and differences of particular technological and stylistic traits through time and over space. This type of analysis is possible for Postclassic slipped pottery because «subtle social relationships through a society's social representations of technology influence the physical action of the material world» (Lemonnier 1992: 23). Thus, through analysis of the Mayas' choices of technologies and styles, it is possible to determine the extent to which social representations are reflected in the development and performance of technological and stylistic action in order to define social identity and ethnic boundaries.

CLAYS OF THE PETÉN LAKES REGION

Lakes in the Petén region are a result of the kartstic landscapes and regional faulting. The underlying sedimentology is a series of limestone, dolomite, shale, and gypsum beds that date from the Cretaceuous to Tertiary periods (Vinson 1962). Lake cores from Cogwill et al. (1966), Cowgill and Hutchinson (1963), Deevey et al. (1979, 1980), and Brenner (1994), provide mineralogical information for the clays that line lakes and are found in surrounding areas. Each lake differs as to its specific water chemistry because of the base exchange of clays, evaporites (especially gypsum), and dolomite (Table 1; Deevey et al.1980: 440).

Table 1. Ionic Proportions in Petén Lake Waters (percentage composition in milliequivalents) (from Deevey et al. 1980: Table 8).

	Ca	Mg	Na	K	HCO_3	Cl	SO_4
Sacnab	24.65	9.35	12.47	4.48	35.6	8.30	5.13
Quexil	36.3	5.7	6.8	1.7	40.6	5.4	3.5
Petenxil	39.7	3.5	5.4	.4	32.8	2.8	15.3
Sacpuy	27.2	8.1	13.5	2.6	37.0	8.7	3.0
Petén Itzá	32.0	14.5	3.8	1.1	14.9	3.1	30.5
Macanché	14.6	28.0	4.8	1.8	19.7	5.6	25.5
Salpetén	28.7	16.7	3.4	.6	.8	2.6	47.1
Yaxhá	32.07	5.64	8.77	3.23	39.5	6.67	4.07

Cowgill and Hutchinson (1963) and Cowgill et al. (1966) cored Lake Petenxil, Bajo de Santa Fé, and Aguada de Santa Vieja. X-ray diffraction (XRD hereafter) analysis of the clay sediments determined that lake beds were dominated by halloysite ($2H_2O$ form) with minor amounts of montmorillonite, gypsum, quartz, and pyrite (Cowgill et al. 1966: 52). However, when cores from the northern portion of the Petenxil drainage were tested by XRD, the clay matrix was almost entirely montmorillonite with small amounts of halloysite and large quantities of quartz (Cowgill et al. 1966: 55). Lower than expected XRD peaks for Bajo Santa Ana Vieja suggest that bajo sediments were primarily halloysite with poorly formed crystals (Cowgill et al 1966: 125).

Bajo de Santa Fé (near Tikal) contained montmorillonite and halloysite clay minerals with small quantities of quartz minerals (Cowgill et al 1966: 25-26). Low XRD peaks for montmorillonite again suggest small, poorly developed crystalline structures. Cowgill et al. (1966: 26) state that clay minerals form poorly developed crystals when sediments are embedded in high concentrations of organic matter and are formed *in situ* by decomposition. In addition to the gray montmorillonite/halloysite clays of Bajo de Santa Fé, Cowgill et al. (1966: 32) also examined local red clays that were predominately a montmorillonite and quartz mixture with little halloysite.

Deevey et al. (1979, 1980) cored the Eocene sediments of Lakes Yaxhá and Sacnab. They stated (Deevey et al. 1980: 301, 420) that the main sediment is allochtonous with a large quantity of montmorillonite clay resulting from the erosion of impure limestone. Brenner (1994) cored lake Salpetén and determined that montmorillonite with a small portion of halloysite lined the lake beds.

From the geological information outlined above, the Petén lakes region appears to be dominated by montmorillonite and halloysite clay minerals. It is likely that the Postclassic Maya potters utilized the local lake clays in their manufacture of pottery because according to Arnold (1985: Table 2.1), most potters will only travel 10 km to obtain clay. As a result of this observation, Maya potters would have used local clays that either came from the lakeshore or near by locations. Because montmorillonite and halloysite co-occur in most places in the Petén lakes region, differences in clay pastes rather than clay mineralogy may serve to better identify technological characteristics necessary for the determination of technological styles.

DATA

The technological styles of Postclassic slipped pottery presented here are based on the Augustine, Paxcamán, Fulano, Trapeché, and Topoxté ceramic groups from the archaeological sites of Ch'ich', Ixlú, and Zacpetén, in Petén and from the archaeological site of Tipuj in Belize. Augustine ceramics (Vitzil Orange-red ware) are composed of a reddish-brown to red clay. This ceramic group has a higher hematite content than the remaining four ceramic groups and has calcite, limestone, quartz, gypsum, chert, chalcedony, and biotite mineral inclusions. The Paxcamán, Fulano, and Trapeché (Volador Dull Slipped/Snail Inclusion ware) ceramic groups consist of a tan to dark gray clay. Calcite, hematite, chert, chalcedony, biotite, and gypsum minerals and shell also occur in the pottery sherds. Topoxté group ceramics (Clemencia Cream Paste ware) are composed of a white to light gray clay. The marly clay paste includes calcite, chert, chalcedony, biotite, hematite, and quartz minerals.

The author conducted several types of analysis in this study: petrographic analysis, limited XRD analysis, and strong acid-extraction inductively coupled plasma spectroscopy (ICPS) analysis. Petrographic analysis identified non-plastic inclusions, minerals, and rock fragments in the sherd paste. In addition to the quantitative analysis (kind and number of minerals present in the sherd paste), their size, distribution, and texture were also determined with the use of published charts (Orton, Tyers, and Vince 1993: 238-239). Although the author was able to identify many minerals and organic matter present in the sherd paste, petrographic analysis is limited because it cannot reveal the full mineralogical composition of a pottery sherd (e.g., the clay minerals). Because of this limitation, petrographic analysis was combined with XRD in order to obtain a fuller mineralogical complement.

The author analyzed 14 sherds by XRD analysis because their firing temperatures were below 400°C. The analysis was limited to a small sample because clay crystalline structures begin to change above 400°C resulting in the ineffectiveness of XRD patterns in detecting clay minerals and because most of the sherds in this sample are fired to temperatures above 550°C. In addition to mineralogical analysis, the author conducted strong acid-extraction ICPS analysis to obtain the chemical characterization of the sherd paste in order to demonstrate possible behavioral practices. This method makes no pretense of testing only the clay, but analyses the chemical composition of the clay and the temper (Burton and Simon 1996: 408). Because different technologies and pottery groups display different technological characteristics, such as clay and temper, variability in chemical signatures can indicate differences in technological choices. Thus, one can obtain results that predict behavioral characteristics because «the variability is a virtue that informs us about technological choices» (Burton and Simon 1996: 406). As a result of the combination of the three types of analysis and a stylistic analysis of the sherds in this sample, the author detected seven technological style groups from the five ceramic groups (Figure 2).

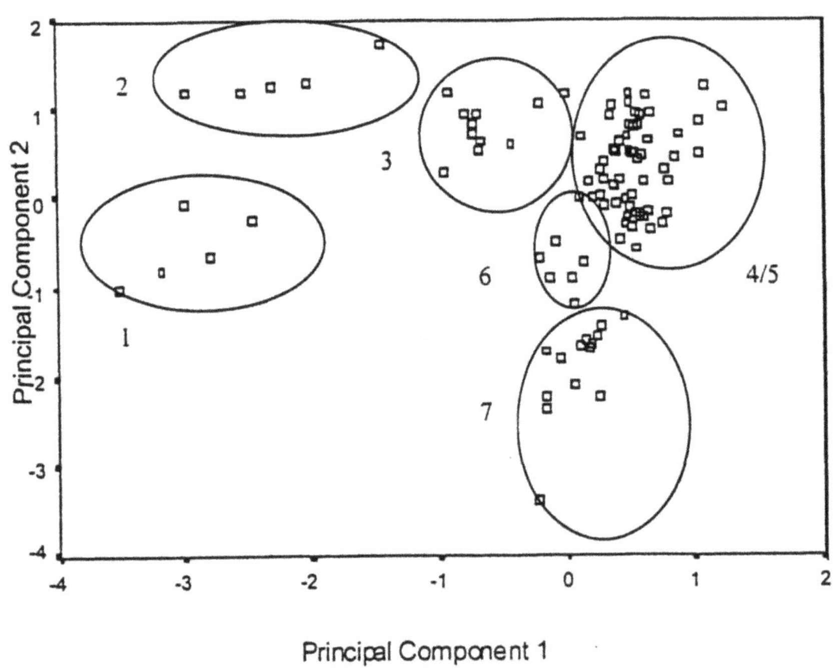

Figure 2. Technological Style Groups.

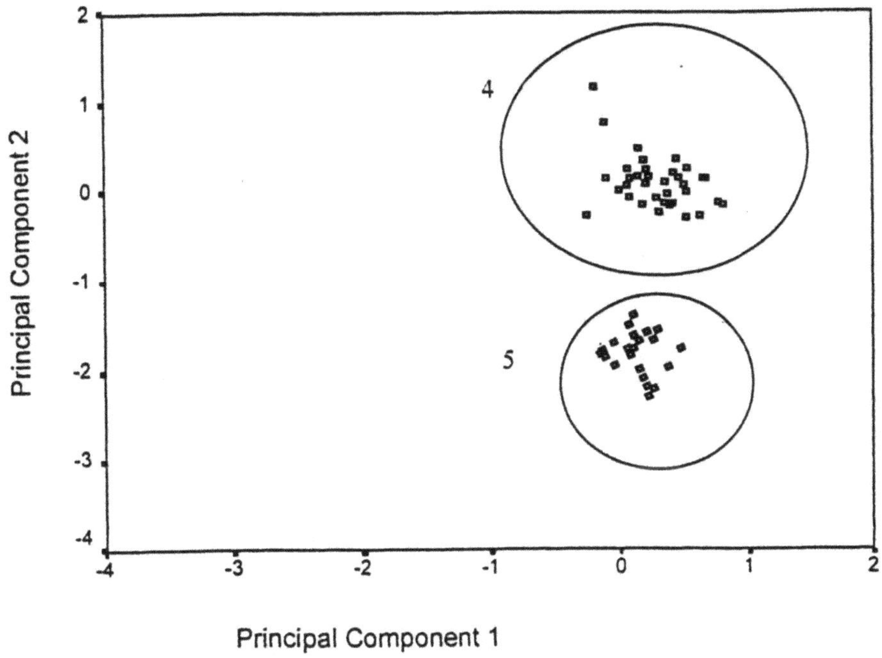

Figure 2 (cont). Technological Style Groups 4 and 5.

Initial technological style groups were detected from a sample of 273 sherds and subsequent visual and microscopic analysis expanded the sample to 551 Postclassic slipped sherds. The ellipses presented here are not ellipses of statistical significance, but ellipses of groupings based on mineralogical and chemical analyses of the sherd paste. The underlying base for the technological style groups comes from principal component analysis of the strong acid-extraction ICPS data. Eigenvalues and associated variances and the rotated component matrix data are provided in Tables 2 and 3, respectively. Mineralogical characteristics match the elemental groupings and are discussed as coherent technological style groups.

Table 2. Principal Eigenvalues and Associated Variance*

Component	Eigen-values	% of Variance	Cumulative %
1	5.017	16.724	16.724
2	4.197	13.990	30.714
3	3.329	11.097	41.811
4	2.684	8.946	50.757
5	2.048	6.828	57.585
6	1.579	5.264	62.848
7	1.290	4.300	67.148
8	1.252	4.174	71.322

* Thirty components account for the total variance.

Table 3. Rotated Component Matrix of the First Eight Principal Components

	1	2	3	4	5	6	7	8
Ca	.881*			.206				
Rb	.859*							
K	-.795*				-.279			
Na	-.781*		.392		.254			
La	.703*				-.228			
Si	-.665*						.218	
Sc	.576*	-.254	.257	.339	.355			
Ti		.935*						
Fe		.929*						
B	-.206	.891*						
V	.346	.819*						
Al	-.393	.798*				-.212		
Cu		.528*	.435			.214	.347	
Be			-.716*					.355
Cd		-.246	.691*					
Mn	-.385		.650*					
Se	.231			.714*				
Pb				-.698*				
Ni			.240	-.516*		.495		
Co	.317	-.315		-.450			-.421	-.221
Mg	.387	-.331		-.398	-.230	.321		
Cr		.221	.243		-.747*			
Sr	.536	.227			.617*		.231	
Ba	-.273	.545			.570			
Sb					.476	.637*		
Zn	-.207					.611*		
Tl		-.445		.272	-.225	.452		
Ag							.852*	
As								.786*
Cs		-.246	-.416	.342		-.375		-.458

* Indicates the elements that constitute the principal component

Technological Style Group 1 (n=19)

Group 1 represents Topoxté pottery from Ixlú and Tipuj. Petrographically, this group is dominated by voids in a sherd paste. A few quartz inclusions exist, but they do not occur in any significant quantity. XRD analysis of sherds from the group demonstrates that the clay matrix is composed of montmorillonite clay minerals (Figure 3). Additional minerals detected by XRD analysis and not by petrographic analysis include gypsum and dolomite. Strong acid-extraction ICPS analysis suggests that the group is distinctive due to its moderate concentrations of Fe (iron) and Ti (titanium), and low concentrations of Ca (calcium) and Zn (zinc). The body sherds of this group are thin and exteriorly slipped with no decoration.

Technological Style Group 2 (n=21)

Group 2 represents Augustine sherds from Ch'ich', Ixlú, and Zacpetén. Petrographically, voids dominate the sherd paste. Small amounts (less than 1%) of quartz, calcite and hematite are also present. XRD analysis indicates that montmorillonite clay minerals occur in the sherds from this group (Figure 4). Strong acid-extraction ICPS analysis distinguishes this group from Technological Style Group 3, also an Augustine ceramic group, because of its low concentrations of Ca (calcium). The majority of the sherds in this group are slipped without decoration (Augustine Red); however, three sherds have either black line painted decoration (Pek Polychrome) or incisions (Hobonmo Incised). Plates and drums are the only forms in this group (Figure 5).

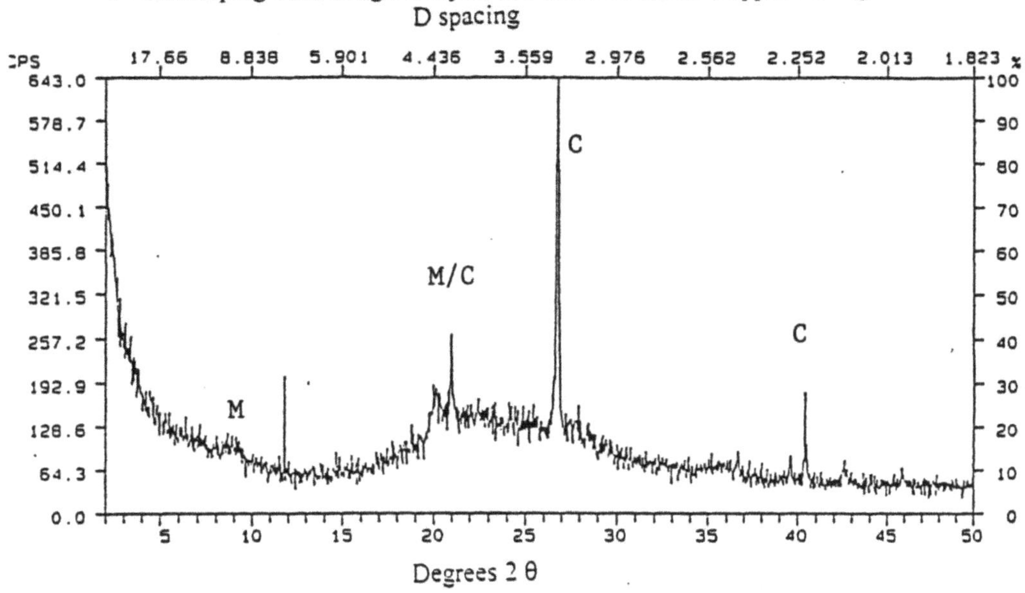

Figure 3. XRD graph of Technological Style Group 1 (M=Montmorillonite and C= Calcite).

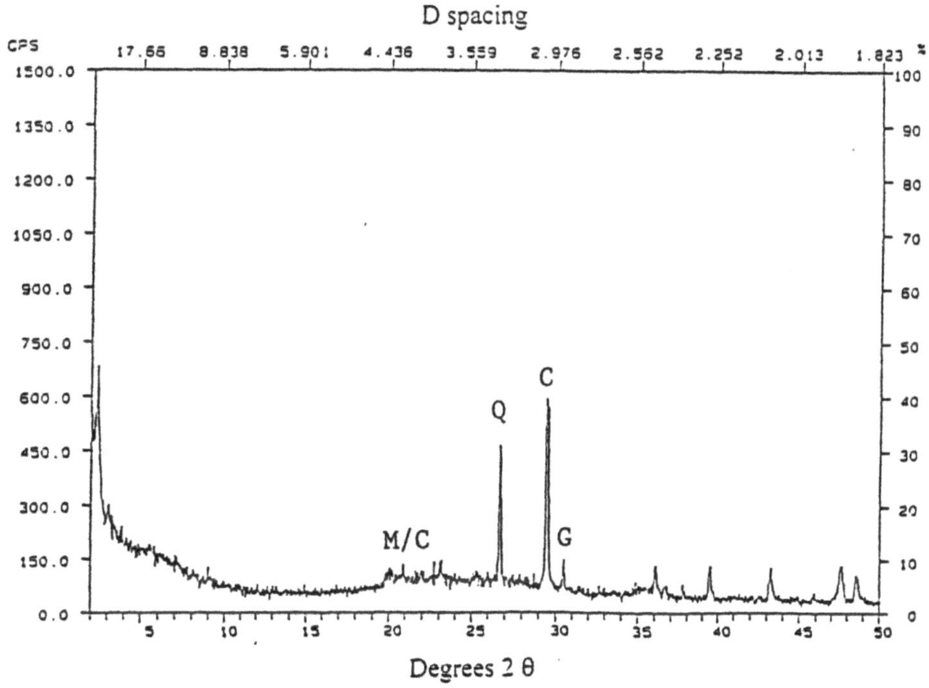

Figure 4. XRD graph of Technological Style Group 2 (M=Montmorillonite, C= Calcite, Q=Quartz, and G=Gypsum).

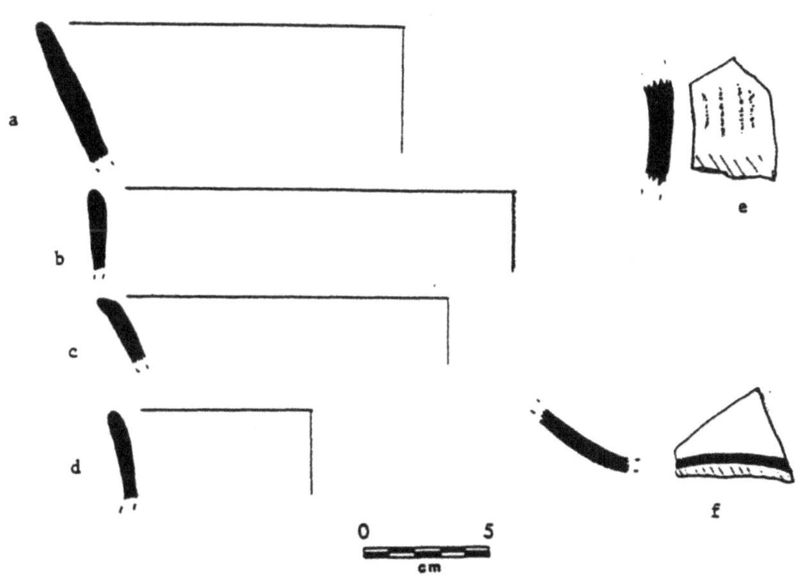

Figure 5. Technological Style Group 2 sherd profiles: Augustine Red (a-d); Hobonmo Incised: Hobonmo Variety (e), and Pek Polychrome (f).

Technological Style Group 3 (n=133)

Group 3 represents Augustine sherds from all four sites in this study. Petrographically, subhedral and euhedral calcite and quartz inclusions dominate the sherd paste with lesser amounts of hematite. XRD analysis demonstrates that montmorillonite clay minerals compose the clay portion of the sherds. In addition to clay minerals, XRD analysis indicates that the clay paste also includes gypsum (Figure 6). Strong acid-extraction ICPS analysis separates Group 3 from Group 2 because of its high concentration of Ca (calcium). Jar, plate, and drum sherds are decorated by incisions (Hobonmo Incised), black (Pek Polychrome), or red and black decoration (Graciela Polychrome). Decorative motifs include plumes, birds, mats, and the *Ilhuitl* glyph (Figure 7).

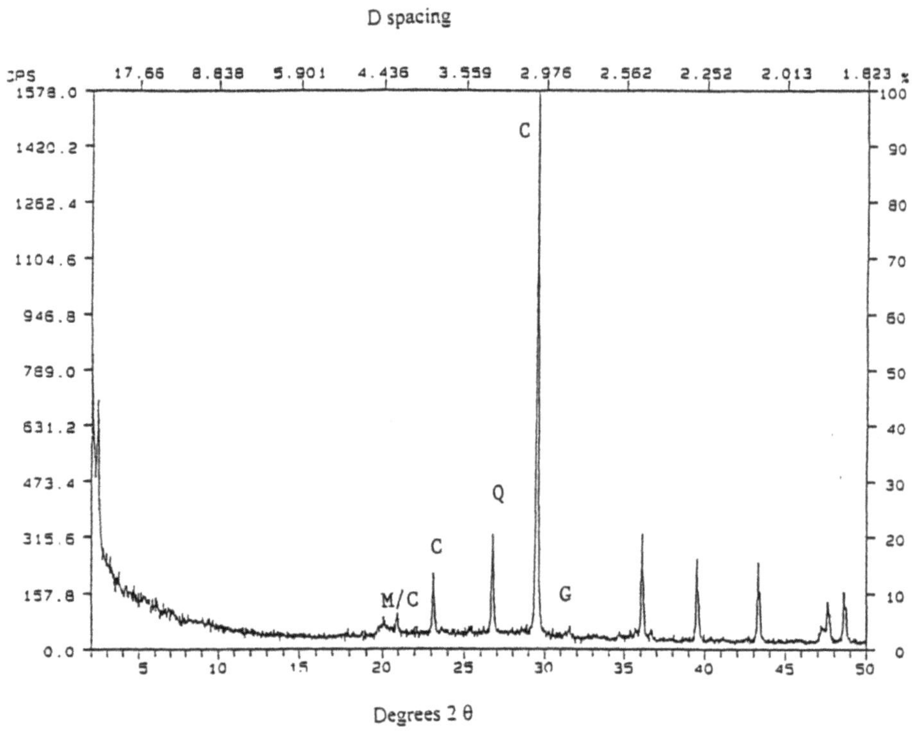

Figure 6. XRD graph of Technological Style Group 3 (M=Montmorillonite, C= Calcite, Q=Quartz, and G=Gypsum).

Figure 7. Technological Style Group 3 sherd profiles: Augustine Red (a-d); Pek Polychrome (e), Hobonmo Incised: Ramsey Variety (f), and Graciela Polychrome (g).

Technological Style Group 4 (n= 173)

The cluster labeled 4/5 in Figure 2 is composed of Paxcamán, Fulano, and Trapeché sherds from the four sites in this study. When this group's ICPS results were analyzed separately, two groups formed (Technological Style Group 4 and 5) and the two groups correspond to two very different technological styles (Figure 2). Petrographically, Group 4 has a calcite rich sherd paste characterized by the presence of subhedral and euhedral calcite, voids, biotite, chalcedony, chert, quartz, and hematite minerals, and shell. XRD analysis of sherds from Group 4 demonstrates that montmorillonite is the primary clay mineral and that gypsum also occurs in the sherd paste (Figure 8). Group 4 XRD analysis corresponds to the XRD analysis of raw gray clay from the lakeshore of Zacpetén (Figure 9). Strong montmorillonite peaks do not occur in all sherds because some sherds were fired above 400°C—the temperature at which the montmorillonite clay mineral structure begins to collapse (Ross and Hendricks 1945: 48-54). Strong acid-extraction ICPS analysis separates Groups 4 and 5 because of lower concentrations of Be (beryllium), Cd (cadmium), and Mn (manganese) in Group 4. Group 4 consists of plates, collared bowls, flanged plates and jars, grater bowls, and drums—in other words, all of the Postclassic forms. Two-thirds of the sherds in the group are decorated with incisions (Picú Incised) or black (Ixpop Polychrome), red (Macanché Red-on-paste), or red and black (Sacá Polychrome) design motifs that are geometric, mat, bird, or *Ilhuitl* designs (Figure 10). The other one-third is slipped without decoration.

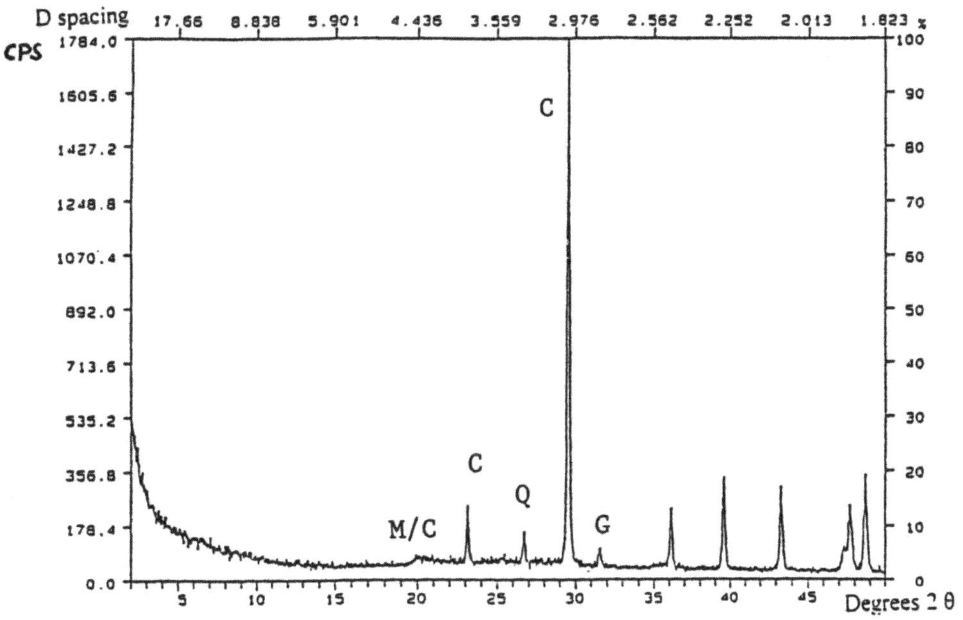

Figure 8. XRD graph of Technological Style Group 4 (M=Montmorillonite, C= Calcite, Q=Quartz, and G=Gypsum).

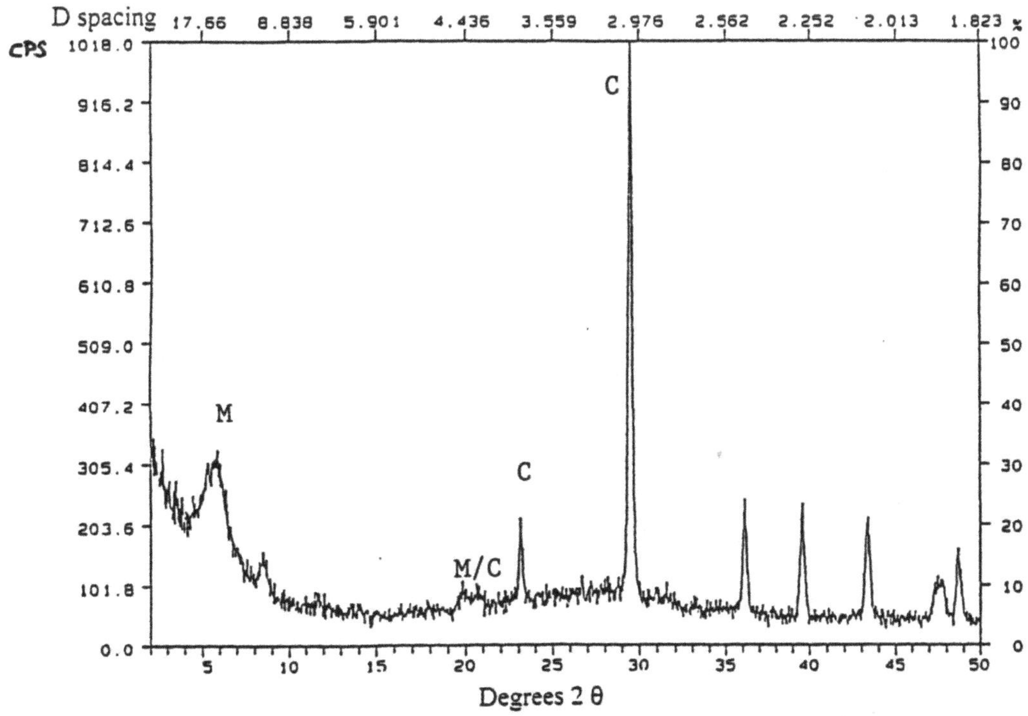

Figure 9. XRD graph of raw clay used for Technological Style Groups 4 and 5 (M=Montmorillonite and C= Calcite).

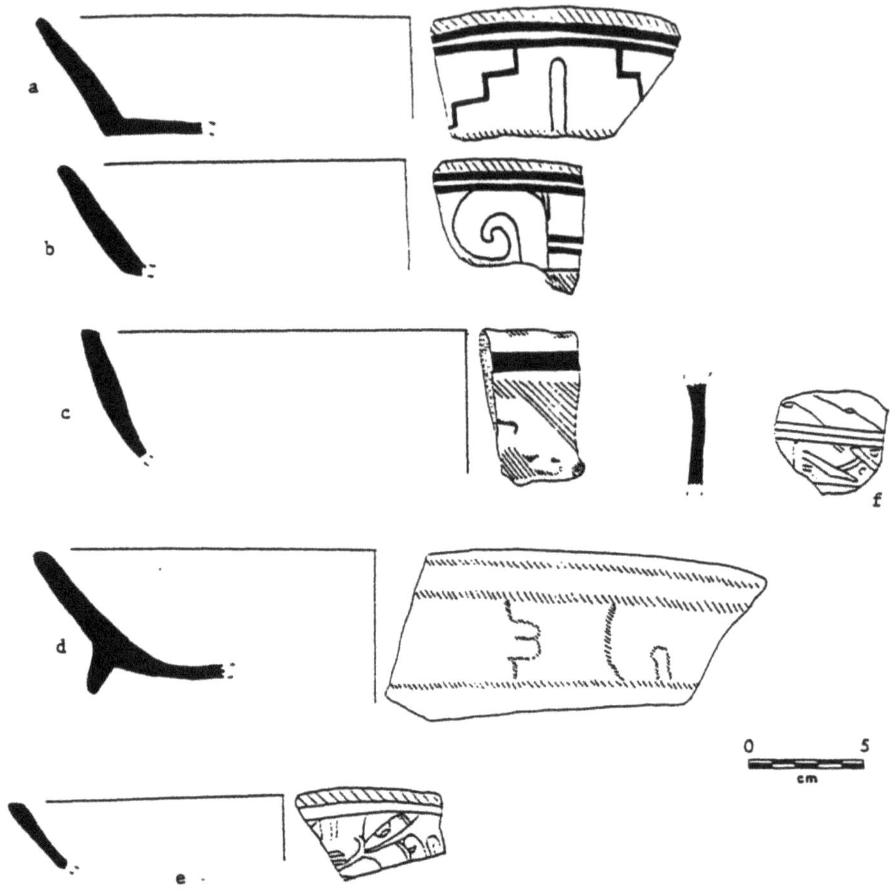

Figure 10. Technological Style Group 4 sherd profiles: Ixpop Polychrome (a-b); Sacá Polychrome (c); Macanché Red-on-paste: Macanché Variety (d); Picú Incised: Picú Variety (e-f).

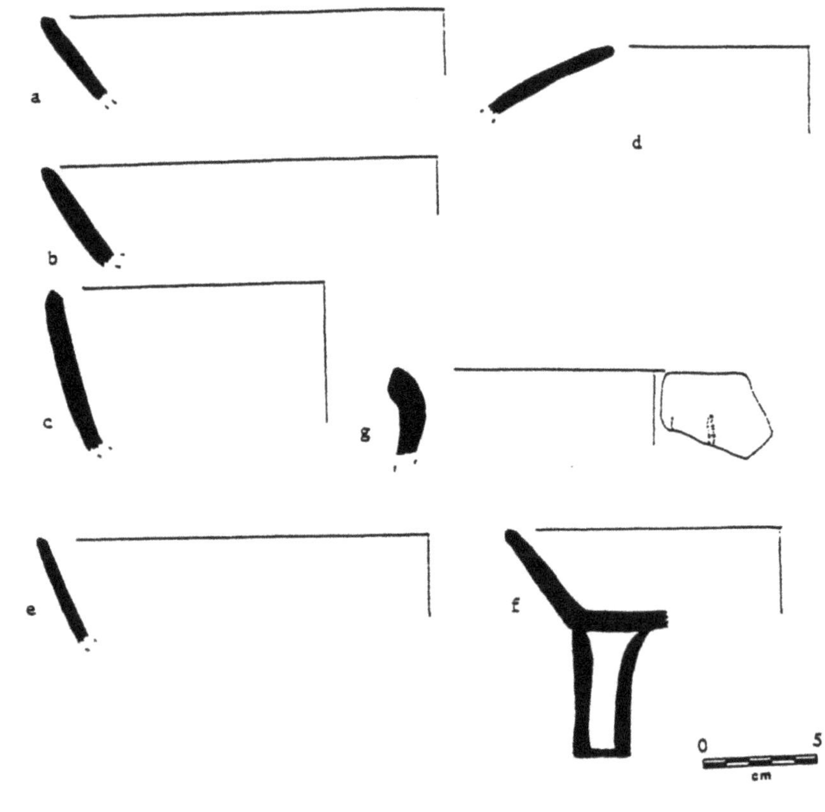

Figure 11. Technological Style Group 5 sherd profiles: Trapeché Pink (a-d); Paxcamán Red (e-f); Xuluc Incised: Tzalam Variety (g).

Technological Style Group 5 (n=101)

Group 5 has a similar mix of Paxcamán, Fulano, and Trapeché sherds from the four sites in this study as does Group 4. Petrographically, the two groups differ because Group 5 lacks euhedral and subhedral calcite, biotite, chert, and chalcedony inclusions, and is instead dominated by cryptocrystalline calcite. XRD analysis of the clay pastes in Group 5 produce graphs similar to those of Group 4 (see Figure 8). Plate and jar sherds that compose Group 5 are slipped but undecorated (Trapeché Pink and Paxcamán Red) (Figure 11).

Technological Style Group 6 (n=37)

Group 6 represents Topoxté sherds from Zacpetén and Tipuj. Petrographically, crytoptocrystalline calcite dominates the groups' sherd pastes. Small quantities (less than 3 %) of subhedral and euhedral calcite, quartz, hematite, and biotite minerals also appear in the sherd pastes. XRD analysis indicates that montmorillonite is the primary clay mineral and that gypsum also occurs in the sherd paste (Figure 12). Strong acid-extraction ICPS analysis distinguishes this group from Groups 1 and 7 (other Topoxté ceramic groups) because of its slightly higher concentrations of Fe (iron) and Ti (titanium). Vessel forms represented by this group include bowls, plates, and jars. The sherds are either slipped with no decoration or decorated with red paint that is darker than the red slip color (Chompoxté Red-on-paste). Curvilinear lines, parentheses, and *Ahau* glyphs encompass the design motifs of the group (Figure 13).

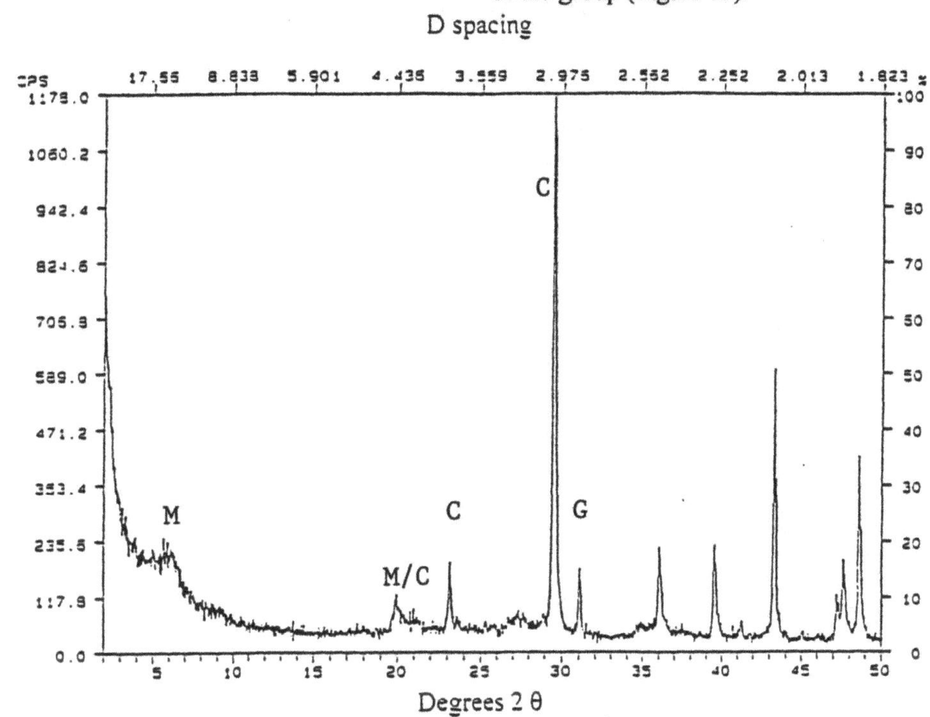

Figure 12. XRD graph of Technological Style Group 6 (M=Montmorillonite, C= Calcite and G=Gypsum).

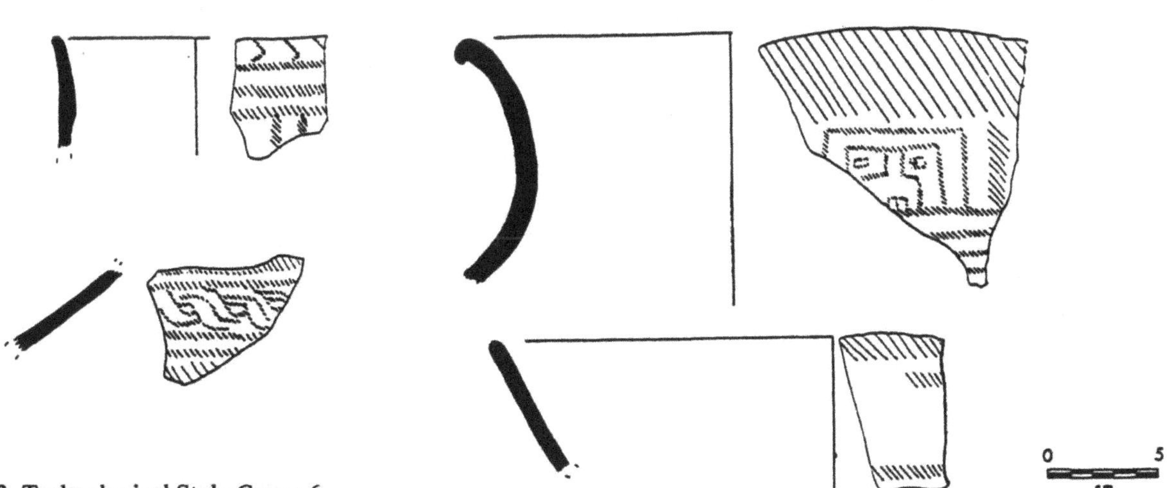

Figure13. Technological Style Group 6 sherd profiles: Chompoxté Red-on-paste: Akalché Variety.

Technological Style Group 7 (n=67)

Group 7 also includes Topoxté ceramic group sherds from Zacpetén and Tipuj. Petrographically, sherd pastes of this group contain subhedral and euhedral calcite, hematite, quartz, biotite, chalcedony, and chert minerals. XRD analysis demonstrates that montmorillonite is the primary clay mineral and that gypsum is also present (Figure14). Plate, bowl, and jar vessel forms occur in this group. Eighty-five percent of the sherds in this group are decorated with red (Chompoxté Red-on-paste), black (Pastel Polychrome), or red and black painted decoration (Canté Polychrome). Unfortunately, design motifs are almost completely eroded (Figure 15).

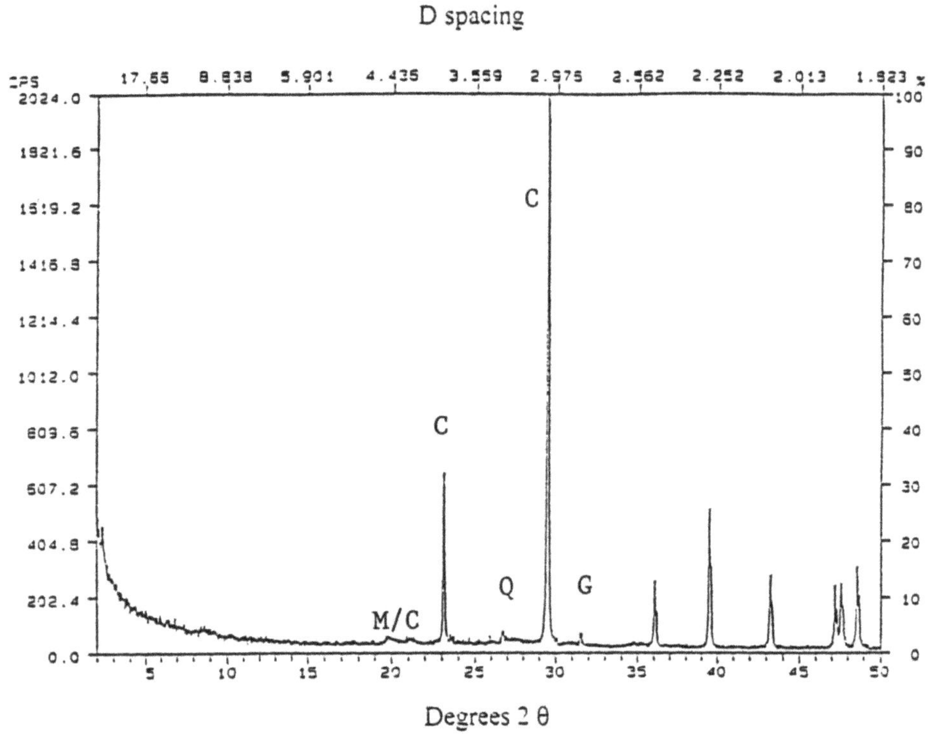

Figure 14. XRD graph of Technological Style Group 7 (M=Montmorillonite, C= Calcite, Q=Quartz, and G=Gypsum).

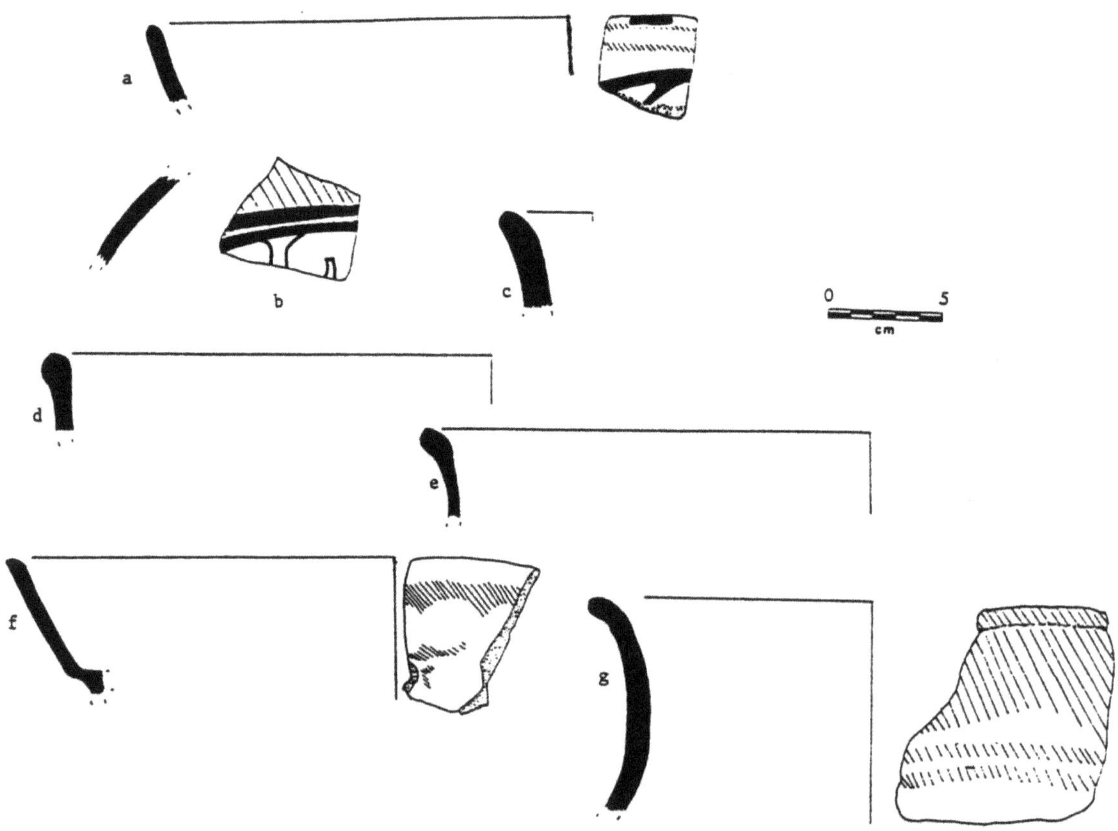

Figure15. Technological Style Group 7 sherd profiles: Canté Polychrome (a); Pastel Polychrome (b); Topoxté Red (c-e); Chompoxté Red-on-paste; Chompoxté Variety (f); and Chompoxté Red-on-paste: Akalché Variety (g).

DISCUSSION

Three attributes distinguished Petén Postclassic slipped technological styles: 1) the sherd paste as a whole (the ceramic ware); 2) the mineral inclusions in the sherd paste; and 3) the decorative or stylistic characteristics of the sherds. These differences, when viewed as coherent combinations of traits, define the Petén Postclassic slipped pottery technological styles that ultimately aid in the identification of social identity.
Because montmorillonite exists in all sherds from the study and occurs locally in every lake bed and inland area of the Petén lakes region, identification based on clay mineralogy alone is not sufficient. Instead, when the author combined clay mineral information with inclusions in the clay paste (attributes 1 and 2, above), technological styles become more apparent. Technological Style Groups 1, 6, and 7 are composed of a white marly clay paste (Clemencia Cream Paste ware) and the groups are further distinguished petrographically by the quantity of voids or the presence of chalcedony and/or biotite. Technological Style Groups 4 and 5 have a tan to gray clay paste (Volador Dull Slipped/Snail Inclusion ware) and are petrographically separated by differences in calcite crystallography and the presence of chert, chalcedony, and biotite. Finally, Technological Style Groups 2 and 3 are characterized by a brownish-red to red clay matrix (Vitzil Orange-red ware) and the groups are petrographically distinguished by the frequency of voids and their presence of chert, chalcedony, and biotite.

The third attribute of technological style groups focuses on stylistic features. Technological Style Group 1 is monochromatically slipped with no decoration, Group 6 has banded red decoration, and Group 7 has banded and unbanded red, black, and red and black decoration. Technological Style Group 4 has decorated types and Group 5 has only monochromatically slipped sherds. Undecorated and decorated sherds from Technological Style Groups 2 and 3 resemble the differences in Technological Style Groups 4 and 5, respectively.

When all of the characteristics discussed above are viewed as coherent groups and compared to other types of material culture in the Petén lakes region, social or ethnic identities become clearer. Technological Style Groups 1, 6, and 7 represent Kowoj social identity. First, Clemencia Cream Paste ware pottery was produced at and traded from the Kowoj site of Topoxté. «[T]he Topoxté Islands do not share in the other Petén Postclassic ceramic traditions. No Trapeché group sherds, for example, were found at Topoxté, and no Chilo Unslipped; only one sherd was tentatively classified as Augustine, and only three sherds were identified as being of probable Yucatecan manufacture. The inhabitants of the Topoxté Islands, in short, seem to have sent some of the pottery throughout a relatively broad territory in Petén [and elsewhere], but to have brought in very little in return» (Rice 1986: 278). Topoxté can be defined as a Kowoj site because Topoxté's architecture (temple assemblages) resembles that of Mayapán (the ancestral homeland of the Kowoj) (Pugh, personal communication 2000). Second, Topoxté pottery is found more frequently at sites within the ethnohistorically defined Kowoj territory.

Kowoj identity may also be defined by red and black and red painted decoration that appears in Technological Style Groups 3, 4, and 7. Two characteristics of material culture in the Maya lowlands support this proposition. First, red and red and black decorated pottery occurs most frequently in the ethnohistorically defined Kowoj territory. Second, temple structures at Zacpetén, Topoxté, Ixlú, and Mayapán (Kowoj sites) have red and red and black painted stucco (Bullard 1970; Pugh, personal communication 2000; D. Rice, personal communication 2000). While the presence of painted stucco may be the result of preservation, painted stucco is not reported at other archaeological sites in the Petén lakes region.

Itzá social identity is more difficult to define through technological style data because most of the excavated sites have been in Kowoj territory. However, Itzá social identity may be reflected through the use of black decorative painting on Vitzil Orange-red ware pottery (Technological Style Groups 2 and 3). These technological style groups occur most commonly in ethnohistorically defined Itzá territories. Further research into Itzá archaeological sites will contribute to the discussion of technological styles that reflect Itzá social identity.

CONCLUSIONS

Unlike most other chapters in this volume, this chapter focuses on the differences in sherd paste inclusions and decoration to differentiate ceramic groups that reflect social identity. The emphasis resulted from the uniformity of clay minerals in the Petén lakes region, the ineffectiveness of the XRD analysis data, and the difficulties in obtaining provenience information. Although elemental differences in the water chemistry of the Petén lakes exist, in order to determine which lake bed the pottery sherds may have come from, the author needed to analyze sherds, lake beds, and lake shore clay chemistry. Because this was out of the scope of the original proposal, the author focused on the clay paste, inclusions, and decorative characteristics to bridge the gap between clay mineralogy and archaeology.

The combination of clay paste characteristics, inclusions, and decorative attributes resulted in seven technological style groups that reflect social/ethnic identity in the Petén lakes region during the 17th century. The existence of technological styles based on the above three criteria demonstrate that the Kowoj and the Itzá made technological and stylistic choices in the manufacturing of pottery that reflected their separate social identity. Different pottery forms were created with different types of clay pastes, inclusions, and decorations and these choices reflect the compatibility of technological and stylistic choices as defined by cultural groups. The differences also reinforce existing social ideology present in other forms of material culture.

Finally, this study demonstrates that analysis of technological styles can answer behavioral and technical questions using methods grounded in clay mineralogy and geology. This is important because archaeologists and clay mineralogists provide the information that is necessary to answer archaeological and anthropological questions. However, the author's use of technological style and methods of analysis demonstrates one of the ways that clay mineralogical methods can be used in archaeological research.

Acknowledgments

This research was made possible by the National Science Foundation Dissertation Improvement Grant (SBR-9816325) and the Sigma Xi Grants-in-Aid of Research. In addition to the two grants, this work is based on excavations by Proyecto Maya Colonial (1994-1999) headed by Dr. Don S. Rice, Dr. Prudence M. Rice, Dr. Grant D. Jones, and Romulo Sánchez Polo. I am grateful for their support and encouragement. Excavation data for Zacpetén and Ch'ich' Structure 188 is provided by Tim Pugh and excavation data for Ixlú is provided by Romulo Sánchez Polo. XRD analysis was guided by Dr. Richard Fifarek and Paul Robinson of the Geology Department at Southern Illinois University and strong acid-extraction analysis was guided by Dr. Bradley Paul of the Mining and Engineering Department of Southern Illinois University. The final version of the manuscript benefited greatly from constructive comments by Isabelle C. Druc and Bruce Velde.

References

Arnold, D. 1971. Ethnomineralogy of Ticul, Yucatan potters: etics and emics. *American Antiquity* 36(1), 20-40.

Arnold, D. 1985. *Ceramic theory and cultural process.* Cambridge: Cambridge University Press.

Bourdieu, P. 1977. *Outline of a theory of practice.* Cambridge: Cambridge University Press.

Brenner, M. 1994. Lakes Salpeten and Quexil, Peten, Guatemala, Central America. In *Global geological record of lake basins: volume 1*, ed. K.E. Gierlowski and K. Kelts. Cambridge: Cambridge University Press, pp. 377-380.

Bullard, W. R. 1970. Topoxté: a postclassic Maya site in Petén, Guatemala. In *Monographs and papers in Maya archaeology*, ed. W. Bullard, Papers of the Peabody Museum of Archaeology and Ethnology No. 61. Cambridge: Harvard University, pp. 245-308.

Burton, J., and Simon, A.W. 1996. A pot is not a rock: A reply to Neff, Glascock, Bishop, and Blackman. *American Antiquity* 61(2): 405-413.

Cortés, H. 1976. *Cartas de relación,* 9th edition. México: Editorial Porrua.

Cowgill, U., and Hutchinson, G. E. 1963. El bajo de Santa Fé. *Tranactions of the American Philosophical Society* 53(7): 1-51.

Cowgill, U.M., Hutchinson, G. E., Goulden, C. E., Patrick, R., Racek, A. A., and Tsukada, M. 1966. The history of Laguna de Petenxil. *Memoirs of the Connecticut Academy of Arts and Sciences* 17: 7-62.

Deevey, E.S., Brenner, M., Flannery, M S., and Yezdani, G. H. 1980. Lakes Yaxha and Sacnab, Peten, Guatemala: limnology and hydrology. *Archiv Für Hydrobiologie* 57(4): 419-460.

Deevey, E.S., Rice, D. S., Rice, P. M., Vaughn, H. H., Brenner, M., Flannery, M. S. 1979. Maya urbanism: impact on a tropical karst environment. *Science* 206, 298-306.

Dietler, M., and I. Herbich. 1998. Habitus, techniques, style: an integrated approach to the social understanding of material culture and boundaries. In *The archaeology of social boundaries*, ed. M.T. Stark. Washington D.C.: Smithsonian Institution Press.

Hodder, I.1977. The distribution of material culture items in the Baringo District, Western Kenya. *Man* 12: 239-269.

Jones, G. D.1998. *The conquest of the last Maya kingdom.* Stanford: Stanford University Press.

Lechtman, H. 1977. Style in technology--some early thoughts. In *Material culture: styles, organization, and dynamics of technology*, ed. H. Lechtman and R. S.Merrill. St. Paul:West Publishing Company, pp. 3-20.

Lechtman, H. 1979. Issues in Andean metallurgy. In *Pre-Columbian metallurgy of South America*, ed. E. P. Benson. Washington D.C.: Dumbarton Oaks, pp. 1-40.

Lechtman, H.1981. Introducción. In *La tecnología en el mundo andino*, ed. H. Lechtman and A. M. Soldi. México:Universidad Nacional Autóma de México, pp. 11-22.

Lemonnier, P. 1992. *Elements for an anthropology of technology.* Anthropological Papers No. 88. Ann Arbor: Museum of Anthropology, University of Michigan.

Lemonnier, P. 1993. Introduction. In *Technological Choices*, ed. P. Lemonnier, London: Routledge, pp. 1-40.

Orton, C., Tyers, P., and Vince, P. 1993. *Pottery in archaeology.* Cambridge: Cambridge University Press.

Rice, D. S., Rice, P. M., Sánchez Polo, R., and Jones, G. D. 1996. *Proyecto Maya-Colonial: geografía política del siglo XVII en el centro del Petén, Guatemala.* Informe Preliminar al Instituto de Antropología e Historia de Guatemala sobre Investigaciones del Campo en los Años 1994 y 1995.

Rice, P. M. 1986. The Peten Postclassic: perspectives from the Central Peten Lakes. In *Late lowland Maya civilization: Classic to Postclassic*, ed. J. A. Sabloff and E.W. Andrews V., Albuquerque: University of New Mexico Press, pp. 251-299.

Rice, P. M. 1987. *Pottery Analysis: A Sourcebook.* Chicago: University of Chicago Press.

Ross, C.S., and Hendricks, S.B. 1945 Minerals of the montmorillonite group. *United States Geological Survey, Professional Paper* 205B, pp. 23-80.

Roys, R.1967. *The book of Chilam Balam of Chumayel.* Norman:University of Oklahoma Press.

Sackett, J. 1985. Style and ethnicity in the Kalahari: A reply to Wiessner. *American Antiquity* 50, 154-159.

Schele, L., and Grube, N. 1995. The proceedings of the Maya hieroglyph workshop: Late Classic and Terminal Classic warfare, March 11-12, 1995.

Thompson, J. E. S. 1951. The Itza of Tayasal, Petén. In *Homenaje al doctor AlfonsoCaso.* México: Imprenta Nuevo Mundo, pp. 389-400.

Vinson, G.L. 1962. Upper Cretaceous and Tertiary stratigraphy of Guatemala. *Bulletin of the American*

Association of Petroleum Geologists 46(4): 425-456.

Wiessner, P. 1983. Style and social information in Kalahari San projectile points. *American Antiquity* 48, 253-276.

CONCLUSIONS

Bruce Velde
Laboratoire de Géologie, École Normale Supérieure, Paris

The different studies presented here indicate some of the great variability which one can expect to see in the future of Archeo-ceramic studies. It seems clear that as time goes on the use of geological and physico-chemical methods and observations are going to be part of the basis of modern archaeological work. The tools at the disposal of the practising archaeologist will increase in number and variety. This prospect indicates that the study of the past by physical artefacts will be very exciting. However such a future means that young people must be better trained in a wider variety of fields than before. The challenge is not for young people but for their mentors.

This being the case, what can one say of the variety of study presented in this symposium volume which represents some of the changes to be seen in the future? Initially it is interesting to note the difference in overall objectives. I believe that one can more or less divide the subject into two groups, one North American in approach and the other European (lacking other contributors for the moment). The European approach is more one of specific observations of objects on a very material level. The questions of how and why remain anchored in the material. One is more interested in the methods, materials and needs of the potter in selecting and producing the ceramic objects discovered by the archaeologist. It is relatively rare that ethnologic considerations enter into the considerations of ceramic research, at least as it is reported in the first instances of publication. Generalisations about the organisation of society come later using different aspects of archaeological record such as buildings, their groupings, agriculture, etc. This might be due to the fact that much research on ceramic materials is done on materials coming from rather well established societies where there are many elements available to consider.

North American studies, in a very general way, tend to try to place the ceramicist in his (her) societal context. Here one not only tries to determine the methods used but more importantly the reasons for a choice of materials, decorations and firing techniques used in producing a ceramic object. These objectives are somewhat different from those common in Europe and certainly more ambitious.

However, both approaches will use the same observational methods in different combinations to gain insight into the production processes of ceramic materials. It is evident that the ceramic study today goes far beyond the traditional use of ceramics as a dating tool. The chemical variability of a type of object, derived from the use of clay resources and additives (temper) as a tool to determine provenance can be used to indicate choices and strategies of individual or groups of potters which respond to specific conditions of supply, custom or innovation. More information is sought and found, as demonstrated in the papers presented above, than has been the case in the past. The study of the specific elements of a ceramic (paste composition, grits, oxidation state, new phases produced and so forth) all points to new data which must be integrated into the corpus of information already accumulated in the fields of ceramic technology, geology, and geomorphology. It is interesting to see how different techniques of observation can be used to gain similar or identical information. Optical petrographic microscope analysis can be supplemented or replaced by X-ray diffraction, and at times bulk chemical analysis. Identification of a specific mineral can be determined by optical methods, or in using the electron microprobe or the SEM. Not only is the presence of a grain important but the shape and abundance can give insight into the sourcing and treatment of materials used for the production of a ceramic paste. The multiplicity of methods of research gives wider latitude to the individual ceramic archaeologist in choice and strategy of study. Also it allows one to use local or available facilities which may not be presented to each researcher.

It is clear that the present trend of study is one of a more intelligent use of resources in the laboratory. The range of questions asked is greater now than in the past and the means of obtaining the answers is continually being enlarged. The future gives exciting prospects.

www.ingramcontent.com/pod-product-compliance
Lightning Source LLC
LaVergne TN
LVHW071219210926
848482LV00015B/851
* 9 7 8 1 8 4 1 7 1 1 7 5 1 *